THE *Field Guide* TO Goats

THE *Field Guide* TO Goats

By Cheryl Kimball, C.V.T.

Voyageur Press

First published in 2009 by Voyageur Press, an imprint of MBI Publishing Company, 400 First Avenue North, Suite 300, Minneapolis, MN 55401 USA

Voyageur Press titles are also available at discounts in bulk quantity for industrial or sales-promotional use. For details write to Special Sales Manager at MBI Publishing Company, 400 First Avenue North, Suite 300, Minneapolis, MN 55401 USA.

To find out more about our books, visit us online at www.voyageurpress.com.

Library of Congress Cataloging-in-Publication Data

Kimball, Cheryl.
The field guide to goats / by Cheryl Kimball.
p. cm.
Includes index.
ISBN 978-0-7603-3522-2 (flexibound)
1. Goats. 2. Goat breeds. I. Title.
SF383.K564 2009
636.3'9–dc22

2008052993

Edited by Danielle Ibister
Designed by Jennie Tischler
Printed in China

Front cover: Cheryl Kimball
Back cover and spine photos: Daniel Johnson
On the frontis: Boer. *Shutterstock*
On the title page left: Nubian. *Shutterstock*
On the title page right: Goats browse next to a barn. *Shutterstock*
Opposite: Boer kid. *David Watts Jr., Dreamstime*
Opposite contents: White goat. *Shutterstock*

Dedication

Dedicated to Thingamabob (1996–2002) and Doohickey (1996–2008), two great goats, together again.

Acknowledgments

Thanks to the American Dairy Goat Association, the Oklahoma State University Breeds of Livestock, and the individual breed associations for allowing me to use their information about breed standards.

Thanks, too, to the following for providing photographs of goats: 7 Fainting Acres; Fran Bishop of Rainbow Spring Acres, Pygora Breeders Association; Michelle Blair of Eagle's Wings Ranch; Jen Brown; Debi Carroll of Ugotabkidn Goat Farm; Phyllis Clayton, Golden Guernsey Goat Society; Copper Creek Boers; Lori A. Corriveau; Gurney Davis; Susan Ducharme; Ernie and Roxanne Gray; Scott Herbolsheimer of Summit Pack Goat; Natascha Jewell of Rainbow Basin Dairy Farm; Daniel Johnson, Paulette Johnson, and Shirley Fernandez of Fox Hill Farm; Paulina Lenting-Smulders; Laurie Macrae of White Sage Farms; Betty Moon of Mystick Acres Farm and Rabbitry; Meghan Murphy, Smithsonian's National Zoo; Kathleen Musser of Blue Ridge Dairy Goat Farm; Mia Nelson of Lookout Point Ranch; New Zealand Dairy Goat Breeders Association; Nicole Riley; Lori Schmick of Sunshine Meadows Farm; Mary Schowe; Ken and Janice Spaulding of Stony Knolls Farm; Susan Stewart; Jennifer Stultz, Gravel-Ends Ranch (www.capriella.com/GravelEnds); Tim Swain of 2T's Farm; Baldur Tryggvason; David Watts Jr.; Sarah L. Zimmerman of Dusty-K-Ranch; and Yvonne Zweede-Tucker of Smoke Ridge Ranch.

I also wish to thank Lori and Peter at Jenness Farm (www.jennessfarm.com) in Nottingham, New Hampshire, for allowing me to take photos at their farm. The owners of Via Lactea Farm (www.vialacteafarm.com) in Brookfield, New Hampshire, also generously let me roam around one afternoon snapping photos. Both of these hardworking farms deserve our support!

Contents

Introduction

There's an old saying about there being two kinds of people in the world: people who have chickens and people who are thinking about getting chickens. I think that for people who live on more than a couple acres of land, this could be said about goats. You either have them or you've always thought about getting a couple.

We brought home a couple of goats thinking they would help us get the brush surrounding our property under control. There was a wether named Thingamabob and a doe named Doohickey. Within a couple months, the goats—which we thought were going to work for their room and board—became just as pampered as our dogs and our horses. To have goats help clear brush, you need to do one of two things: get a herd large enough that they could actually make a dent just in sheer numbers (which in turn would require a goat herder to tend them or miles and miles of fencing) or get just a couple and figure out how to concentrate their area, moving them when they've done the right amount of damage to a small spot.

The key issue with the latter approach is figuring out how to concentrate them. Because you are innately dealing with

These goats lead a charmed life at the New Hampshire Farm Museum, where they entertain visitors. The two dark goats are African pygmies. The white goat is probably a Boer/Nubian cross; he was found as a kid on the side of the road. *Cheryl Kimball*

Opposite: The Nubian is a friendly, popular breed for dairy farms and as pets. *Shutterstock*

unruly brush—the reason the goats came in the first place—fencing those areas you want cleared is difficult at best. We tried tying the wether and letting the doe loose, figuring she wouldn't leave the wether far behind. This tactic resulted in the wether "choking himself down"; a goat's windpipe is located near the front of the throat, and it is easy for air to be cut off. I heard the doe bleating frantically and found the wether passed out in the bushes, having gotten his tether all tangled up in the brush and around his neck. I managed to untangle him, and he resurrected. The result was that they never cleared another piece of brush—and that Thingamabob fell so in love with me as his rescuer that I could sit in a chair and read and he would just come over and rest his chin on top of my head.

Goats get under your skin. When my Oberhasli wether Sprocket looks at me with those yellow fishlike eyes, his one-horned head tilted to one side and the pink tip of his tongue sticking out between his lips, I can't help but laugh. Doohickey would drive me crazy every three weeks or so bleating miserably for a couple days to let the whole world know she was in heat, but I loved her anyway, bought special grain to make sure her aging body was getting the nutrition it needed, and brought her a bucket of warm water twice a day for eight months out of the year.

Caretaking for a couple of pet goats isn't much work in the general scheme of livestock husbandry. If you start off with the right fencing, all you need to do is provide them with some good hay and fresh water, vaccinate them once a year, and trim their hooves every few months and they are set. They do well in a small yard. Creating a working dairy herd is a different story; while rewarding, this requires real dedication, as you will see in chapter 5 of this book.

Friends of mine have kept goats in all sorts of different situations: pets that roam freely during the day and spend the night in a horse stall; dairy herds that spend nights in the barn and days outside, lining up at the parlor door every twelve hours begging to be milked; companion goats that live in the same paddock with their horse friends 24-7. Goats will adapt to almost any living arrangement so long as it includes shelter, food, and companionship.

Goats also seem to replenish themselves endlessly. When Thingamabob died four years after we brought him home, within two weeks I had a new companion for the pining Doohickey. They just kind of do that to you.

If you have been thinking about getting goats, you are not alone. And you will be joining the ranks of those who find goats to be simply entertaining.

Opposite: A pampered Toggenburg doe spends the day comfortably in her paddock.
Daniel Johnson

Chapter 1

Goat Evolution, History, Research, and Lore

Where did goats come from? History tells us that the Middle East is the birthplace of the caprine species, with their origin focused in the Euphrates River Valley, Turkey, and the Zagros Mountains of Iran.

Goats have the distinction of being among the first domesticated animals around ten thousand years ago (new research shows that dogs may have preceded them) in the Middle East region known as the Fertile Crescent, a swath of exceptionally rich soil considered to be the cradle of civilization. Goats are perhaps the first wild herbivore that became domesticated, before cows or sheep, likely due to their manageable size and friendly nature.

Goat remains have been found at archaeological sites in western Asia, such as Jericho, Choga, Mami, and Cayonu, which dates the domestication of goats to between 6000 and 7000 BC. Their ancestry is fairly clear. The major genetic contributor of modern goats is the Bezoar goat, which ranges from the mountains of Asia Minor across the Middle East to Sind. DNA research shows that today's goats descended from a relatively small number of animals.

Goats may be easily tamed, but they are independent animals. Unlike domestic sheep, domestic goats easily revert to feral or wild conditions given a chance. In fact, the only domestic species that will return to a wild state as rapidly as a goat is the house cat.

There are currently more than 700 million domestic goats in the world, across three hundred different breeds, although just a few of those breeds are recognized by breed registries or are well represented. Only .2 percent of the world's goats reside in the United States; most are in the deserts of Asia and Africa. Goats have

The ibex in the Neger Desert of Israel may be an ancestor of today's domestic goat. *Shutterstock*

Opposite: A wild goat finds a good rock for the best view. Goats are skilled at climbing. *Shutterstock*

been thought of as the poor man's cow, because they are small in size, can thrive on meager food sources, and can cope with harsh environments. Useful traits indeed, especially in desert environs!

Goats Do Roam

"Goats do Roam" is a brand of wine, believe it or not, but in this case goats roaming refers to how goats managed to travel around the world. For reasons similar to why goats are called the poor man's cow, early explorers often brought goats aboard their ships. These small, hardy animals could provide plenty of meat, milk, and hide. Like horses, they often ended up as feral herds on deserted islands wherever these oceangoing travelers went. In many instances, goats were released on islands to forage and fend for themselves, then picked up on the return voyage to provide fresh meat or milk for homebound vessels.

Goats sailed aboard the *Mayflower* when it came to North America in 1620. Documents show that cattle were the primary livestock transported on the vessel, but two "she goats" were included among the holdings divided up among Plymouth Plantation settlers.

Goats made an appearance in North America upon disembarking from the *Mayflower*, a replica of which can be found in Plymouth, Massachusetts. *Shutterstock*

In 1904, St. Louis hosted the World's Fair and, along with it, the nation's first dairy goat show. Exhibitors imported new goat breeds from around the globe for the event. Among the newly debuted animals were Alpine dairy goats from the Black Forest of Germany. The premier exposition sparked the formation of the American Dairy Goat Association (ADGA), which today maintains the largest goat registry in the United States.

Goat Organizations

Many organizations are available to help goat keepers with their animals. Almost all of the breed entries in chapter 3 are accompanied by the breed association, as well as the organization's web address. The Internet is an exceptionally useful tool for general and unusual goat information, since goats are considered a minor farm animal in most of the world, and there is not a lot published about goats elsewhere.

Here is a list of some of the more useful general goat organizations that you will want to check out.

American Dairy Goat Association (ADGA)

www.adga.org

The ADGA sets the standard for eight recognized dairy breeds: Alpine, Saanen, Sable, Oberhasli, Nubian, LaMancha, Toggenburg, and Nigerian Dwarf. However, the ADGA also provides great information and resources for anyone who owns a goat of any breed or mixed breed. The organization collects data on goat milk production; it also sanctions goat shows and keeps records of performance. If you are raising one

A pair of Alpine dairy goats were exhibited at the 1904 St. Louis World's Fair, although those goats are now lost to history. Modern Alpines in the United States descend from goats imported in the 1920s. *Daniel Johnson*

of the ADGA's recognized breeds on any significant scale, you definitely want to be a member.

International Goat Association (IGA)

www.iga-goatworld.org

Self-proclaimed as the only international association dedicated to the development and promotion of goat activities, this organization spans sixty-three countries. It helps organize an international goat conference every four years. It endorses research centers that work with goats. Its website archives its newsletters, provides information on goats as sustainable agriculture, and, in conjunction with Reed Elsevier, publishes the online version of the *Small Ruminant Research* journal. The IGA site also hosts a blog and maintains a calendar of goat-related conferences. This organization is for the serious goat person.

American Goat Society (AGS)

www.americangoatsociety.com

This organization was founded in 1936 to promote the U.S. dairy goat industry. It provides breed standards, information on showing, and National Dairy Herd Information Association (DHIA) testing and milk production awards.

American Meat Goat Association (AMGA)

www.meatgoats.com

This organization, focused on the meat goat industry, had a heavy emphasis on Texas members when it first started in 1992 but now boasts members in nearly forty states as well as Canada. Its long list of admirable goals includes promoting goat meat as an acceptable and healthy meat sold to both direct consumers and grocery stores; educating the general public on the benefits of eating goat meat; promoting meat goats as a show ring class; and educating potential meat goat raisers.

East Texas Goat Raisers Association (ETGRA)

www.etgra.com

Established in 1996, the East Texas Goat Raisers Association is dedicated to the promotion of the goat industry in East Texas and elsewhere. These folks are serious and well organized. The ETGRA sponsors shows and has a youth program. The site is full of information designed to promote raising goats. Despite the regional focus, it is a great information resource for anyone interested in goats.

American Sheep and Goat Center (ASGC)

www.sheepandgoatsusa.org

Formerly the National Sheep Industry Improvement Center, this organization is dedicated to promoting profitable small-ruminant farming. It works toward providing loans through the Sheep and Goat Industry Grant Initiative.

National Livestock Producers Association (NLPA)

www.nlpa.org

Although the NLPA is predominantly cattle-oriented, its website does include information and news about all livestock.

National Institute of Animal Agriculture (NIAA)

www.animalagriculture.org

The mission statement of the NIAA is to be the "forum for building consensus and advancing solutions for animal agriculture and to provide continuing education and communications linkages to animal agriculture professionals." Its website contains an archive of recent news articles and educational resources for all livestock breeders, including a backlog of *The Sheep and Goat Health Report* newsletter.

A stunning mountain goat. *Shutterstock*

Goat Equipment Suppliers

Numerous businesses sell supplies specifically for goats. A few of these follow.

Hoegger Goat Supply

www.hoeggergoatsupply.com

You can get everything at Hoegger, from nursing bottle nipples for $1.85 to milking systems starting at $1,500; from artificial insemination equipment to insect control. Hoegger, which has been in business since 1935, has made it their business to supply all things goat. You can be sure if you get it from Hoegger, it is as goat-specific as you can get. This is one any goat farmer will bookmark on her or his computer.

Caprine Supply

www.caprinesupply.com

Another all-around goat-supply business, at Caprine Supply you can find cheesemaking equipment, books and magazines, and almost anything goat-related.

A Mongolian goat herder tends his herd from horseback. *Shutterstock*

Goat Supplies and Services

www.goatsuppliesandservices.com

This general goat-supply business includes fecal-testing supplies as well as herbs and natural products for goats, such as blends to boost the immune system or to help with digestion.

Jeffers Livestock

www.jefferspet.com

This general livestock and pet supplier carries goat-specific supplies, such as fence chargers, antibiotics, vaccines, supplements, and halters.

Owyhee Packgoat Supplies

www.owyheepackgoatsupplies.com

This interesting company carries supplies needed to use goats as pack animals. Products include halters, leads, saddles, and bags, as well as materials to stake goats out and dog deterrents.

Northwest Pack Goats & Supplies

www.northwestpackgoats.com

This company manufactures its own pack goat equipment and supplies. Its online store carries saddles, pads, collars, leads, coats, first-aid kits, and everything else you need for the unique experience of camping with pack goats.

Goat Research

Although not as widespread in the research arena as many animals are, goats are the subject of more research than one might think.

University of California, Davis

www.ucdavis.edu

UC Davis hosts a dairy goat research facility established in 1980 and used to teach students about dairy (and meat) goats. UC Davis is developing transgenic goats with genetic antibacterial properties that express in milk. According to facility manager Jan Carlson, these genetic properties are "beneficial to human and animal health, offer some mastitis protection, and positively impact the processing properties and shelf life of milk." The research facility and its herd of around seventy-five goats provide animal science students significant hands-on experience.

Langston University

www.luresext.edu

The E (Kika) de la Garza Institute for Goat Research at Langston University in Oklahoma has a stated mission to "develop and transfer enhanced goat production system technologies, with impacts at local, state, regional, national, and international levels." The research farm comprises 320 acres of pasture, a creamery, and a goat herd that ranges from 800 to 1,400 goats. The institute provides outreach and education, as well as annual events such as Goat Day. Langston's Research and Extension programs also have a strong international focus, with projects going on all over the world.

Sheep & Goat Research Journal

www.sheepusa.org

This journal, put out by the American Sheep Industry Association, focuses predominantly on sheep, but each issue typically includes an article or two about goat research. The research papers mainly discuss nutrition, feeding, and grazing concerns.

Goat Lore

Goats have been well represented in folklore and literature. Perhaps the best-known fairy tale involving goats is Three Billy Goats Gruff. If you recall, to reach a choice pasture, three billy goats have to cross a bridge under which lives a troll. As each goat tramps across the bridge, the troll threatens to eat him. The first two sacrifice the goat to follow as bigger and therefore, they each explain, the troll would get more food. The greedy troll waits for the third goat, which gores the troll with his

The five-goat statue in Guangzhou City, Guangdong Province, China, is a symbol of the city. *Shutterstock*

A beautiful herd of goats roams the Corsican mountains. *Shutterstock*

horns and tramples him. And all three goats go off to their destination, a lush field where they get fat. This version of the tale is of Norwegian descent. Other European versions exist in which the troll is a wolf.

The Chinese Year of the Goat occurs every twelve years. Those who were born under the goat sign are thought to be artistic, kind, gentle, intelligent, and sensitive. On the negative side, they also tend to be self-indulgent, insecure, and ungrateful! Their element is earth, their season summer, and their compass point south-southwest. Famous goats are actress Nicole Kidman, businessman Bill Gates, actor Lawrence Olivier, and writer Jane Austen.

Capricorn, symbolized by the goat, is the tenth sign of the zodiac in Western astrology. Capricorns have birthdates between December 22 and January 19. According to astrology.com, Capricorns are "industrious, efficient, organized, and won't make a lot of waves." That's pretty much the goat! Capricorns are extremely dedicated to their goals and typically take a businesslike approach to almost everything they do.

Goats in Books

Heidi, written by Johanna Spyri in 1880, is the story of an orphan girl who moves to the Alps to live with her cranky, aloof grandfather. Heidi befriends a young goat-herder boy and

Goats populate many fairy tales. *Shutterstock*

comes to love the mountains and the goats that inhabit them.

Beatrice's Goat is a great children's book about a family in Uganda that is given a goat from the Heifer Project (see below). The goat provides the family with milk, bears two kids that they sell at market, and literally changes the lives of Beatrice and her family.

Goats even show up in contemporary fiction—the novel *Zoology* by Ben Dolnick is about a young man who works at a children's zoo and decides to let his favorite goat run loose.

Goat Books

Books a goat keeper will want to have on the reference shelf include the following:

The Merck Veterinary Manual. Also available online in searchable format and free, this 1,000-plus-page manual is very clinical, but most find it indispensable.

Keeping Livestock Healthy by N. Bruce Haynes, D.V.M. A general livestock care book that covers all aspects of all livestock species, including goats.

Storey's Guide to Raising Dairy Goats: Breeds, Care, Dairying by Jerry Belanger. This book has been around for awhile and revised a couple of times. While it leaves out a lot, what is there is useful.

Goats: Small-scale Herding for Pleasure and Profit by Sue Weaver. Many of the goat books focus on dairy herds, but for the small-scale owner, this is a good book.

Raising Meat Goats for Profit by Gail B. Bowman. Anyone raising goats for meat will want to read a book or two specific to that goal.

The Goat Handbook by Ulrich Jaudas. This book looks at goats as pets and is good for those who own pet goats.

The Heifer Project International

The Heifer Project is a nonprofit organization dedicated to addressing world hunger. Its approach is to provide families in need with livestock such as goats,

A herd of goats stays together in the Himalayas. *Shutterstock*

Scientific Classification of the Domestic Goat (*Capra Hircus*)

Phylum: mammalia (the young are born alive and nurse on milk)
Order: artiodactyla (even-toed, hoofed)
Family: bovidae (ruminants with hollow horns that don't shed)
Genus: *Capra* (genus includes only goats)
Species: *hircus* (domestic goat)

Shutterstock

cows, sheep, rabbits, and chickens. The livestock comes with education about husbandry and sustainable agriculture. A single goat can provide a family with a daily supply of fresh milk; it can also provide offspring to sell or provide the family with meat or establish a larger dairy herd to supply others with milk. To learn more about Heifer International and to donate a goat to a family, check out www.heifer.org.

Basic Goat Facts—and Myths

True goat lovers take offense at the terms "nanny" and "billy." As with deer, the correct term for an adult female is "doe" and the term for a breeding male is "buck." A castrated male is called a "wether."

Also offensive to goat aficionados is the false idea that goats will eat anything, including tin cans. This could not be fur-

A LaMancha samples a flag before discarding it as nonfood. *Shirley Fernandez, Paulette Johnson*

ther from the truth. Goats are discerning eaters. They may try anything, but they will only eat things that are appealing to them. Any goat owner knows there is no "five-second rule" when it comes to goats. If a piece of hay touches the ground, it is unlikely it will be eaten by a goat; it may be trampled and slept on, but it is no longer considered food!

Other Goat Lore

Goats have figured in many strange places in lore and legend. Here are a few.

The Billy Goat Tavern

The famed Billy Goat Tavern in Chicago was allegedly so named when a goat fell off a truck and wandered into the tavern, according to the official tavern website (www.billygoattavern.com). Tavern owner William Sianis grew a goatee to mimic the animal and was nicknamed "Billy Goat" from then on. In 1945, he tried to use the goat as a good-luck charm for the Cubs, but they wouldn't allow the goat into the ballpark. He cursed the Cubs to never win the World Series until the goat was allowed into the park. That year, the Cubs lost their bid for the World Series pennant; this loss was blamed on the "Billy Goat curse." It is said that in 1974, Billy Goat Sianis' nephew, accompanied by a goat, drove to the Cubs ballpark in a limousine and the goat was again denied entrance. The Cubs had a losing season. When the team finally began letting the goat into the park, the Cubs flourished. When the championship game was played elsewhere—and the goat was left behind in Chicago—the Cubs lost. This pattern repeated itself for many seasons. Cubs fans are now known to chant "Let the Goat In!" to try to influence their team to win.

The Sound of Music

Remember the puppet show in the film *The Sound of Music*? Maria outdoes herself helping the captain's children put together a puppet show for their father's new fiancée. The theme of the spirited

An old boy takes a nap in the shade. *Shutterstock*

Hilton Goat RIP

According to celebrity gossip in August 2006, Paris Hilton bought a cemetery plot to bury her pet goat, Billy. The plot is in the Pierce Brothers Westwood Village Memorial Park in Hollywood, right next to the gravesite of Marilyn Monroe.

performance is a goat herder and his goats each falling in love. The puppet show ends with the goat herder marrying the girl and the buck and doe goat rubbing cheeks, happily in love. This fabulous scene is the final straw for the fiancée, who clearly will never live up to Maria's ability to entertain the children.

The Legend of the Dancing Goats

According to legend, coffee beans were discovered by a goat herder who fell asleep and awoke to find his herd missing. When he located his goats, they were dancing capriciously beside a thicket of bushes covered with red berries. The goat herder was hungry and decided to partake of the berries, and he too began dancing. A monk wandered by and saw the dancing goats and goat herder. The monk later tried the berries himself and found that he was awake most of the night. He decided to share the energetic fruit with others. This legend led to the Dancing Goats coffee and espresso bars, as well as Wandering Goats coffee.

A boy goat herder in the Kalahari Desert of Africa herds his goats home. *Shutterstock*

Chapter 2

Goat Practicalities

For the most part, goats are hardy animals. They require a minimal amount of care: basic shelter, quality hay and/or browse, and water at all times. Goats do require a couple of annual vaccinations and are susceptible to a few common diseases that you will have to watch out for, all of which are discussed later in this chapter. But first let's get to some basics.

Acquiring a Goat

Before you need to worry about housing, feeding, and breeding, you will need to acquire some goats. How do you go about that?

Part of your goat acquisition research depends on what you have planned for your goats. If you are looking for a couple of pets or a companion animal for a draft horse, you can probably find a goat or two at a livestock animal shelter. These animals have typically been vetted and, if they were malnourished or ill, brought back to good health before they are allowed to be adopted. They have probably also received basic vaccinations. You may be asked to pay an adoption fee

Both male and female goats can have beards and/or horns. *Shutterstock*

Opposite: A kid plays on a tree stump in its goat yard. *Shutterstock*

to cover the cost of the care and feeding the animal has received.

Animal shelters are great places to acquire a pet goat, but if you are looking for more than entertainment and companionship, you need to be more selective in your goat acquisition process. Whether you will be using your goats for meat, dairy, or fiber—and especially if you plan to sell any of those three products—the process for acquiring the foundation of your herd is fundamentally the same.

Foundation Stock

The best way to acquire goats for your caprine business is to visit goat farms that have the kind of goats you want. Don't limit your search to the goat dairy around the corner; goats are small enough to be relatively easy to transport, so you can expand your search for goats without adding much in the way of expense.

With that in mind, select three or more herds to visit, depending on how much time you have to look around. You can find them online or through ag magazines and newsletters. For example, the Department of Agriculture in New Hampshire puts out a newsletter called *The Weekly Market Bulletin*, which includes ads for different species of livestock. Also look at ads for caprine products to identify potential sources of your foundation stock.

Take care to purchase quality goats for your foundation stock, whether you plan to raise a meat, dairy, or fiber herd. This Angora goat is kept for fiber. *Sarah L. Zimmerman, PhD, Dusty-K-Ranch*

Call farmers to make an appointment to visit their farms. Be sensitive to their time limitations. Dairies, for example, will be pretty busy at the beginning and end of the day; around lunch or mid-evening can be a good time to call many farmers. Leave a message if you can and allow them to call you back at their convenience. If they have advertised animals for sale, they will be eager to talk with you.

Go look specifically at the animals for sale and their overall operation in general. While the animals for sale may look and be healthy, the rest of the herd should be just as healthy. Later in this chapter, we will cover common diseases and ailments to be concerned with; review these carefully and ask the farmer about their history with these goat illnesses.

If you are planning to start up a dairy herd, be sure to look at the dairy operation's production records, at least for the line of goats that are for sale. Milk production of the dam is often indicative of production of her offspring. While you can't be guaranteed equal results, these figures at least give you something to base your decision on.

When you respond to ads with animals for sale, keep in mind that you will not have much time to make a decision. This doesn't mean you should make a rash decision, but you do need to realize that if you are planning to visit three meat-goat producers, the first one you go to may not be interested in holding the goats until you visit the other two and make your decision. Be sure to tell the farmer how long you plan to spend making your selection. But don't be pressured: These first animals will be the foundation of your whole operation.

Adding to the Herd via Purchasing

Adding new animals to an existing goat herd makes it take on a different complexion, even if that herd consists of only two goats. When choosing the animals to buy, first examine the weaknesses of your existing breeding stock and select new animals to improve those weaknesses. Try to add animals with a great milk-production genetic line or a history of multiple births. This approach is especially important when choosing the appropriate buck to breed to your does, which will be covered in the next section.

A newborn kid gets his sea legs. *Shutterstock*

A doe and her kid graze in a field. Kids mimic their mother and begin eating solid food within a few days. *Shutterstock*

Adding to the Herd via Breeding

If yours is a dairy operation, you will automatically be adding to the herd via breeding in order to get your does lactating. However, this does not mean you keep all the offspring; in fact, usually it is quite the opposite. Male offspring in a dairy operation typically go to the meat market; they may also be sold as pets or companion animals. Few males are worth keeping as breeding stock; they need to be extremely high-quality animals and, in any case, they need to be biologically unrelated to most of your breeding does. If your dairy doe produces a higher-quality buckling, he may be raised beyond market weight and sold as a breeding buck to another farm.

Most dairy operations will give doe kids a chance to prove themselves as good breeders and milkers. If a doe doesn't prove herself after one or two kidding seasons, she is culled from the herd.

Meat operations also breed in order to expand their herd, keeping only the most well-conformed doe kids for breeding candidates.

Transportation

The method you choose to bring your new goats home will depend on what animals you decide to purchase. If you decide to start with just a few breeding stock (you could get two pregnant does and be well on your way to herd development!), you can easily transport them in a pickup truck. Two kid goats can go in dog crates and ride in the back seat of the truck. Two adult goats will need to ride in the bed; they might even fit in large dog crates, depending on their size. Better still is a truck with a cap over the bed that allows the goats to go loose and avoid the windy ride. Remember to put down a rubber mat and some bedding and they will be quite comfortable. For a ride of more than an hour or so, clip a bucket

Twins are common in goats, as are two at the milk bar. *Shutterstock*

of water to one of the truck bed tiedowns and offer the goats a little choice hay to help keep them calm.

You can even fly goats long distances on commercial airlines. They will need to go in a dog crate. Like all animals being transported by air, you will want to be sure they are well marked to identify their ownership, especially if they get separated from their crate. Be sure the crate latch secures in a way that can't be easily opened. Make sure the crate has a water bottle. (Although the goats may not understand how to use a water bottle, you can show it to them and hope for the best.) Put a small amount of hay in the crate and tape a bag of goat chow to the top of the crate so they can be fed a meal if the flight is delayed for any length of time.

Incorporating a New Goat into an Existing Herd

When your new goat arrives, you don't want to just put it in with your existing herd. Set up an isolation area where the new acquisition can be comfortable for at least ten days, a place where the goat can't touch noses with any others. If you acquired more than one goat from the same place, they can be boarded together in isolation; if you acquired multiple goats from different places, isolate them from your herd and from each other.

A new goat at the hay bunk can cause major disruption in the herd. New animals should be quarantined and then introduced carefully and under supervision. *Shutterstock*

(This means not transporting them together in the back of the pickup!) Ten days is usually enough time for any illnesses to show up in your new stock; isolating them from the rest of your herd during this time will prevent the spread of any potential illnesses. Of course, you should only buy a goat that has been fully vaccinated before you buy it, but new ailments show up all the time and you can never be too cautious about spreading disease.

Feeding

The folklore about goats is that they will eat anything. Not true! Goats will try almost anything, but they are quite picky about what they will actually consume.

Before outlining what to feed your goats, here is some information on how the goat digestive system works.

Goat Digestion

Like sheep, cows, and deer, goats fall into the category of "ruminant." A ruminant has four stomach compartments: the rumen, the reticulum, the omasum, and the abomasum. Each plays an important role in ruminant digestion. This stomach structure makes the ruminant's eating habits considerably different from other livestock such as horses and pigs.

Horses are designed to graze and digest on the move all day long. In contrast, ruminants eat until they are stuffed, then go lie down and digest what

A goat consumes food, then lies down to regurgitate and chew its cud over and over again. This process is called "ruminating." *Shutterstock*

they ate. Digestion begins in the rumen. The rumen is filled with lots of bacteria that aid in primary digestion.

Once the feed is digested in the rumen, it passes in and out of the second part of the stomach, the reticulum. The solid parts are regurgitated in the form of a cud. As the goat lies down and digests, it continuously regurgitates, chews, and reswallows this cud, working all the food it took in through that second digestive process.

The third part of the ruminant stomach is the omasum. This is where water and inorganic elements are absorbed into the bloodstream.

The last chamber is the abomasum, where non-ruminant-specific (monogastric) digestion takes place, moving digesta through the system to the small and large intestines. This final step is the most similar to human digestion.

Bacteria: Good and Bad

The bacteria in the rumen is critical to digestion, but it can also be the cause of problems. When too much feed is consumed or the bacterial flora in the rumen is otherwise disrupted, overactivity of the bacteria causes a serious issue with gas buildup and bloat, which can quickly become fatal. The digestive system in most animals is a miraculous, delicate balance that works great most of the time. When making changes to your goats' diet, always add new foods gradually and under supervision.

Shutterstock

Anyone taking care of goats will discover that they love to eat. What do they like to eat? Here are some guidelines.

Grain

Grain is a must for the milking doe. To produce the optimum amount of milk, a dairy goat needs high-quality hay and supplemental grain. Most livestock-feed manufacturers produce a grain with the accurate protein percentage and other vitamin and mineral requirments essential to the milking doe. A lot of your feed decisions will depend on the feed manufacturers available in your area. Farmers in the Northeast will find grain mixtures that are corn based, while in the West, grains might be more barley based. In the South, rice bran may find its way into the grain base.

In the dairy herd, kids are normally taken from their mother at birth; typically, the kids stay with the mother just long enough to receive the initial milking, which is called colostrum and is rich in antibodies and minerals. Kids are bottle-fed for several weeks but then require grain if they are to meet a market or breeding weight within a practical amount of time. Doe kids not fed grain will probably need another year of growth to reach the average eighty-pound breeding weight. These decisions are based mostly on economics: If you have to wait to breed a doe, you also have to wait to get milk from her.

Grain is beneficial to male goats for several reasons. First, males are susceptible to a buildup of minerals in the urinary tract called urinary calculi, which in turn causes urinary blockage. This condition is potentially fatal and may require surgery. To help prevent this buildup of calculi, male goats should be fed a grain with ammonium chloride in it, which helps balance out the pH of the goat's urine. This applies to both breeding bucks and wethers kept as pets.

Breeding bucks are fed extra grain during the fall rutting season. The extra grain helps them maintain their weight during this extraordinarily active period.

If wethers are being raised as meat goats, grain will help bring them to market weight faster and improve their meat. Male kids intended for market probably won't live long enough for urinary calculi to develop, so you don't need to worry about whether the grain you feed them is supplemented with ammonium chloride.

Minerals and Supplements

Providing your herd with loose minerals and salt blocks allows goats to self-regulate their mineral intake. Livestock are excellent at consuming minerals when they need them.

Another common supplement given free choice to goats is baking soda. Bicarbonate of soda helps neutralize the

Goat grain comes pelleted or textured. Grain is fed to meat goats to bring them to market weight and to dairy goats for milk production. Salt and mineral blocks, loose minerals, and baking soda free choice are supplements you should add to your goats' diet. *Cheryl Kimball*

very actively fermenting ruminant stomach. Again, the goats will eat it as they feel the need.

Hay

Hay is perhaps the most important feed for the confined goat. Dairy herds will do best on at least some portion of high-protein alfalfa hay. For the backyard pet goat, straight grass hay is fine. Wethers, in particular, may experience adverse effects from the large amounts of calcium in alfalfa hay, so neutered males kept as pets should just graze in a pasture, browse the shrubbery, and/or be fed grass hay.

Hay is best—and most easily—fed free choice, but goats waste a lot of hay by picking out the very best parts and leaving the rest on the ground. Once a piece of hay touches the ground, it is not palatable to most goats. Using hay feeders appropriate to goats is the best prevention against wasting excessive amounts of hay.

Hay racks attached to the wall can help; hay that does drop becomes bedding. A traditional feeder is a wooden hay rack with openings that have a round hole to accommodate a goat's head, known as keyhole feeders. Feed stores carry various kinds of metal and rubber racks. If the feed store near you doesn't have what you want in stock, ask to see its catalogs and have something special-ordered.

Hay is a fundamental part of your goat herd's diet. *Shutterstock*

Grazing and Browsing

Goats certainly will graze in a pasture; they can even be turned out with horses and other livestock. But browsing is what goats do best—from rambling through the shrubs to pick at this and that, to standing up on their hind legs and reaching as high as they can for that tasty leaf. Although even a small flock of goats can be effective at brush clearing, you need to have some sort of portable fencing (see later section on fencing) to concentrate their efforts—or you need an extremely large flock.

You might be surprised to find that goats can and will safely eat amazing things like poison ivy and prickly berry bushes. However, to graze/browse your goats safely, you do need to be aware of the types of plants that are poisonous to goats (see page 36) before letting them loose on an overgrown field.

Like deer, goats are primarily browsers, not grazers. To graze is to work over the lawn like a lawnmower; to browse is to pick and choose the choicest, most tender greens wherever the eater can find them. *Shutterstock*

The word "ruminate" is used to refer to being thoughtful and contemplative. Goats can be very thoughtful looking when they are ruminating!

Goats browse in a heather field. *Shutterstock*

Poisonous Plants

Goats are selective eaters, and their browsing habits mean they eat a little of this, then a little of that, moving all the time, never ingesting a lot of one thing. This feeding style means they often do not eat enough of a poisonous plant to cause serious problems. However, you can't rely on this selectiveness: It is the goat caretaker's responsibility to make sure that the pasture is free of plants that are toxic to the goat.

Some plants are toxic only in certain forms, such as the red maple, whose leaves are toxic when they dry in the fall or when they wilt and die after a branch falls off. Keep tabs on your pastures and look for things like fallen branches and remove them when you see them, just to be sure.

Dried leaves are like potato chips to a goat, but it's up to you to make sure there are not poisonous leaves within your goat's confines. *Shutterstock*

Some signs of poisoning include the following symptons:

- Frothing at the mouth
- Teeth grinding
- Bloat
- Colic
- Diarrhea
- Rapid pulse
- Weakness

Treating toxicity from plant poisoning depends on the toxin. Call your veterinarian immediately if you suspect your goat has ingested a poisonous plant and is exhibiting signs of poisoning.

Look in your favorite plant identification book or online to find illustrations and photographs of these plants.

Black Walnut

This plant has been known to kill animals even when it accidentally ends up in bedding. When you buy shavings and sawdust, make certain you are getting dry pinewood byproduct, with no other types of trees mixed in.

Rhododendron

Many common landscaping plants are poisonous to animals, and rhododendron is one of them. Azaleas and yew are also landscaping ornamentals that are very poisonous.

Many common landscaping plants are toxic to goats, including the rhododendron. *Cheryl Kimball*

Common Milkweed

The leaves, fruits, and stems of milkweed are all poisonous to goats.

It may be the plant of choice for monarch butterfly caterpillars, but milkweed is poisonous to goats. *Cheryl Kimball*

Mountain Laurel

This ubiquitous woody evergreen plant is also known as "lamb's kill" for its poisonous trait.

Bracken Fern

There are many different types of ferns, but these are the poisonous ones.

Buttercups

All parts of this lovely yellow-flowered plant are poisonous.

Beautiful buttercups are toxic to most livestock; grazing animals usually avoid them. *Cheryl Kimball*

Housing

Goats are extremely hardy, and while they do require housing, it need only be minimal. Basically, they need a shelter to get out of precipitation and wind, preferably with flooring to get them up off the damp ground. The shelter does not need to be complicated or fancy in any way.

Different parts of the country will have different requirements, depending on the climate. In the North, they will need some ability to get warm; if there are several goats, body heat will help, so it is best to give them a shelter that is at least partially enclosed to help trap that body heat. Don't enclose them too much, however. Moist, ammonia-laden warm air is bad for the respiratory system.

Some sort of bedding is necessary to absorb urine and feces. (Sorry, they won't accommodate you by doing this all outside.) Bedding also helps to prevent them from lying in their own waste. If you feed your goats hay inside the shelter, you may find

Housing can be simple or more deluxe, like this prefab shed. *Cheryl Kimball*

that they waste enough to make their own bedding without you having to add any.

If you do add bedding, where you live in the country will probably have a bearing on what you use; straw, shavings, and sawdust are all common beddings. Although you don't need to clean a goat shed every day, it is important to make sure that your goats aren't lying in wet bedding.

Make sure the housing for your goats is well-ventilated. *Susan Ducharme*

Straw is a good bedding type for goats. It is especially good for birthing, since it isn't dusty and won't cling to a wet newborn like sawdust. *Shutterstock*

Fencing

Fencing in goats is like herding cats: It can seem all but impossible. It is best to do it right the first time rather than dealing with goats that have learned to escape. Don't be duped into thinking that because goats are not big or heavy (compared to a horse or a cow), their fencing doesn't need to be sturdy. It does! The same agility that allows goats to stand on their hind legs and browse high up in trees is also used to climb on fences—and all but the sturdiest fence comes crashing down. Fugitive goats can decimate a neighbor's garden in very little time.

There are several options for fencing in goats. A lot will depend on your landscape, barnyard, budget, and the goats themselves.

Hog Panels

Hog panels, typically sixteen feet long and four to five feet high, are made of welded heavy-gauge steel rod that creates a mesh pattern of four-inch squares. Hog panels tend to be sturdier than stock panels and have extra horizontal rods at the bottom. The bottom rods prevent piglets from getting through and adult hogs from reaching through to the greener grass on the other side, which stresses fencing. Hog panels can be secured to either wood or metal fence posts. If you decide to use stock panel, plan to make your posts closer to better stabilize the panels.

Wood Fencing

Plain old post-and-board fencing is definitely an option for goats. Be sure to use four cross rails, as goats are small enough to squeeze underneath the typical post-and-board construction used for horses—and you also want to be sure to keep predators out. To that end, you can line the lower part of your board fence with mesh fencing. Board fencing is attractive, but it can be an expensive option and requires considerable maintenance.

Electric Fencing

Goats will respect electric fencing, which can be the cheapest fencing option of all and certainly is an appropriate choice for large acreage. However, you must check your electric fencing often; your goats will, and they must never find it lacking!

If you have a small herd that you would like to move around from pasture to pasture, you can use portable electric mesh fencing that rolls out and rolls up with ease, along with a portable solar-powered fence charger. This mesh fencing is also useful for pasture management to close off sections of pasture and open up others.

Fencing goats is an art. This fence is a sturdy combination of wood and hog panel. *Shutterstock*

Health Care

Domestic goats do need basic health care and maintenance—vaccinations, hoof care, and horn issues, to name a few. Although goats are extremely hardy and you probably won't deal with much in the way of illness, there are some bodily issues you will learn more about than you want to. And there are a few special concerns regarding health care for pregnant does, which we cover later in this chapter.

One thing that needs to be said: It seems like each year there are fewer and fewer large-animal veterinarians. And finding veterinarians who have any experience with or knowledge about goats is even more unusual.

As a goat owner, you will need to do a lot of your own veterinary care. You will need to learn how to administer shots, both for routine vaccinations and, in the case of illness, to administer medications like antibiotics. This is just the way it is.

Find yourself a network of goat owners to trade thoughts with. There are several active goat listservs on the Internet. But be cautious about the medical advice you get. Some of it will be very helpful and accurate; some of it will not. It's impossible to understand your exact situation without the other person actually being there. It's not that these lists don't serve a useful purpose; they do. But keep your skeptic hat on, and think things through yourself.

Large-Animal Veterinarians

Large-animal veterinarians are in scarce supply across many parts of the country. Companion-animal veterinary practices are thriving, and more and more veterinary students have made their way to the relative comfort and higher wages of caring for people's house pets. Large-animal veterinary practice means a lot of slogging through mud and rain in the dark of night (animals rarely get seriously ill during daylight hours), doing veterinary work in the stall of a barn, and having to manhandle large, unruly animals who are in pain and don't appreciate shots and sutures. You can see where the decision to go into small-animal practice is appealing.

The moral of the story? Be prepared to do much of your goat veterinary care yourself, but locate a large-animal veterinarian and develop a relationship with him or her before you need emergency veterinary services beyond the scope of your knowledge and abilities. This may still mean having to transport the animal to the veterinarian, so figure out how you might do that well before the need arises.

Paulette Johnson

Don't underestimate your own knowledge. You will soon become an expert in your own right.

Needles and Syringes

Learning how to give shots is relatively simple. The hardest part will be restraining the goat. It helps to have a second person hold the animal for you. The dairy goat is so accustomed to being in the milk stand that often you can restrain her there. Give her a little grain to occupy her, and have your helper make sure she doesn't startle and jump off the stand when you administer the injection. Chances are she won't even notice.

You can order sterile plastic syringes and needles online. Lots of one hundred are the cheapest way to go. You can buy syringes with needles attached or syringes and needles separately. Both have their advantages. The attached needles mean you don't have to think about anything; everything is right there, ready to go when you open up the package. However, you may want to have two or three different-sized syringes. Vaccines are often just a couple cc, but another common injection you will give is antibiotics and these require a larger dose. In this case, buying lots of one hundred may not make as much sense.

As for needles, you will want 16-, 18-, or 20-gauge needles for goats. The larger the number, the finer the needle. Antibiotics tend to be thick, so you may want to fill the syringe with a larger-sized needle than you will use for the goat. Keep this in mind if you buy needles already attached to syringes, as it can be difficult to swap needles on these.

A 20-gauge needle will probably be sufficient for almost every injection. Sometimes you just need to do this a bit to determine your own preferences.

Giving Injections

There are two main ways that you will give shots: intramuscular (IM) and subcutaneous (SQ or "sub-Q"). An intramuscular shot, by definition, goes into the muscle. In the goat, this is typically administered in the flank or in the shoulder muscle.

Subcutaneous injections are given under the skin. The technique is to pull up a "tent" of skin, insert the needle in the side of the tent at a 45-degree angle, release the tent, and push the plunger.

Before pushing the plunger, however, on either SQ or IM injections, you always want to pull the plunger just a little to see if you get blood back. If you do, pull the needle part way out and reposition it. You generally do not want to inject medications into blood vessels; at the least they can sting, and at the worst they can cause greater harm, even death.

There is a third injection type, and that is administering medication into the jugular or drawing blood from the jugular for testing. You are likely to leave this up to the veterinarian, although once you have a few years of experience under your belt, this will probably become part of your repertoire of skills as well.

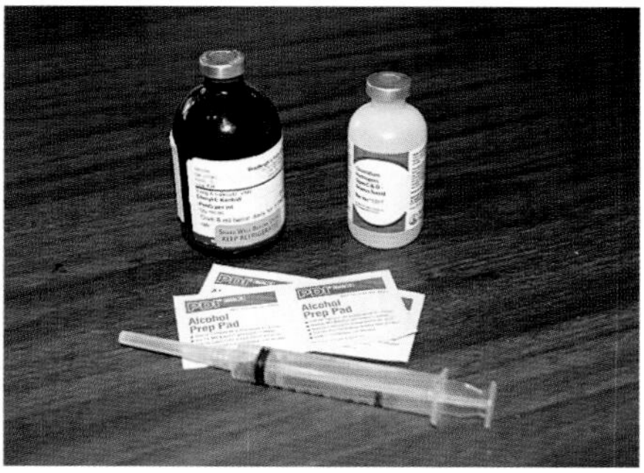

Needles, syringes, and antibiotics are good to have on hand for basic goat care. *Cheryl Kimball*

Common Vaccinations

For the adult goat, the two main vaccinations that goat caretakers should concern themselves with are CD&T and rabies. Selenium is an issue in some regions.

CD&T

CD&T is actually one shot that vaccinates against two common goat diseases—clostridium (C and D types, D being the one that most often affects goats) and tetanus. Although goats have only a modest susceptibility to tetanus, it is enough of a risk to recommend an annual vaccine.

Clostridium is a bacteria normally present in the soil and, therefore, in the stomachs of grazing animals. It becomes a problem when it proliferates, which often happens in the spring when animals that have been locked in the barnyard eating dry hay all winter go out and gorge themselves on grass. Not only are they ingesting grass, they are also ingesting an overabundance of the clostridium bacteria, to toxic levels. And the overpacked stomach gives the bacteria the reduced-oxygen environment that it prefers. The organisms then multiply rapidly and produce toxins that cause digestive tract problems. In otherwise healthy animals, it usually just causes intestinal problems, but in young, old, or compromised animals, death can occur.

Rabies

All mammals are susceptible to rabies. All livestock should be vaccinated every year, as they are often exposed to wildlife that commonly carry rabies, such as skunks, foxes, and raccoons. When purchasing these vaccinations from one of the many livestock supply houses online, be sure you are getting a brand of rabies vaccine that is specifically for goats.

Selenium

Parts of the United States, particularly the Northeast, are deficient in the mineral selenium. This deficiency can cause what is known as white muscle disease; kids are especially affected and exhibit stiffness in the hind legs and quite often die. To help avoid this, does are injected with a selenium and vitamin E supplement within two to four weeks of kidding. The kids are injected at three to four weeks of age.

Goat Vaccinations

Many vaccinations are not tested on goats because, compared to other livestock, there is not a high enough population to make the research worthwhile. However, since sheep once populated a large part of the United States, there are plenty of vaccines approved for sheep. Using those vaccines for the other small ruminant, goats, is typically fine, although it is always worth asking your veterinarian.

Diseases

Goats are not susceptible to too many diseases, but there are a few ailments on their short list of problems that any goat owner should know about.

Bloat

Bloat is caused by overactive bacteria in the rumen. The bacteria activity creates a buildup of gases, causing bloat—evidenced by a noticeably bloated stomach and probably some signs of discomfort. Some goat owners keep the over-the-counter medication Gas-X on hand to give to goats they suspect are bloated. Baking soda can also help reduce gas buildup. In severe cases, the side of the goat needs to be incised to let the gases escape. If the gas buildup does not successfully subside, it can cause the animal's rumen to burst, which is fatal.

CAE

Caprine arthritis encephalitis (CAE) is a highly contagious viral disease all goat raisers want to avoid. Once clinical signs appear, it is most often a death sentence. The affected goat simply wastes away. Goats can be carriers of CAE without ever showing signs of the disease themselves, which makes it particularly challenging to prevent. Intense herd management can help you have a CAE-free, or "clean," herd.

Johne's Disease

A disease known well in the bovine dairy industry, Johne's disease is well controlled. A wasting disease like CAE, Johne's affects young animals, although infected kids can be older than a year old before showing signs. Tests used in cattle often come up negative in goats, even though on autopsy—the only reliable form of confirmation of the disease—the same animal tests positive. Control of outside animals and keeping a clean herd is the only effective method to prevent this disease.

Mastitis

Inflammation of the mammary glands is a disease to which any dairy animal is susceptible. Be sure to practice intense hygiene when milking. Look for early signs of mastitis (off-color milk, inflamed teats/udders), and treat immediately. Antibiotics and topical medications are some forms of treatment.

Parasites

Like all livestock, goats need to be systematically treated for internal parasites. Goats are susceptible to a long list of worms. You need to check with your veterinarian to learn what worms are a problem in your part of the country, then medicate with the proper dewormer at the right lifecycle of the particular worms to be sure you are treating appropriately. Also, perform regular fecal tests on your herd, so that instead of guessing, you know exactly what you are dealing with. Various dewormers are available in injectable or oral forms. It is cheap to treat for worms and expensive not to.

External parasites include lice, mange, and screw worms (the latter particularly in the South). A well-managed, healthy herd will seldom encounter these external parasites.

Hoof Care

Unless your goats spend most of their days on rocky or hard terrain, you will need to keep their hooves trimmed. This is something you can do yourself; there should be no need to call anyone in or have the veterinarian do it. If you don't trim your goats' hooves, the feet will become deformed or suffer from hoof rot, causing discomfort for the animal.

Go to one of the goat supplier websites (see chapter 1) and order a pair of hoof trimmers. The orange-handled ones are inexpensive and work perfectly.

Wear gloves when trimming hooves. Hoof trimmers are lethal and goat hooves themselves can be sharp, especially when a goat new to hoof trimming is flailing his leg around, ready to take a slice out of your arm.

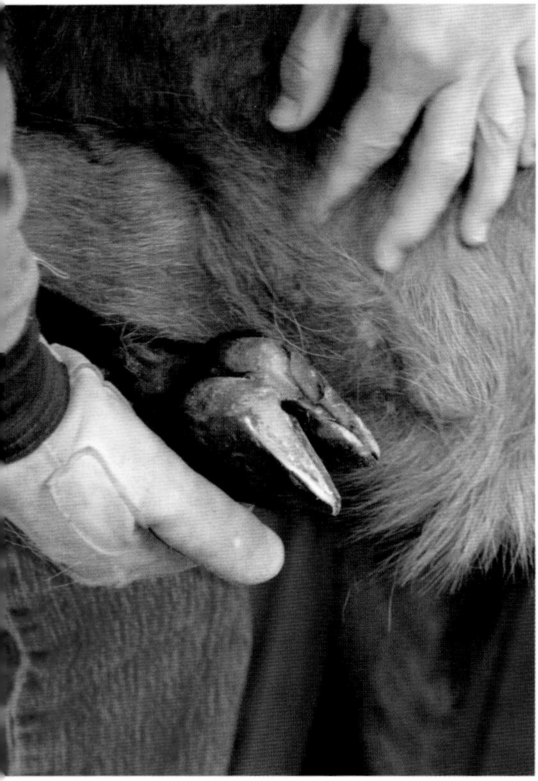

Goats have cloven hooves, with each part known as a "claw." This left claw has been trimmed; the right claw still has a flap of overgrown hoof. *Cheryl Kimball*

You can use a goat stand, which allows you to stand upright while trimming. Or you can tie the goat securely to a post and push his body up against a wall to secure him while you pick up a foot. Most goats that are accustomed to having their foot picked up will be calm and gentle about it. Goats who are not yet used to this activity may struggle; just hold the goat's leg firmly until she or he stops struggling. Then start your task. If she struggles again, hold and wait again. The first few times you work with a goat on foot trimming, you might want to do two feet one day and two the next.

Start by picking the dirt out from the trapped areas where the hoof wall has overgrown. Dirt gets caught under this flap of hoof and can cause hoof rot. Back hooves tend to grow faster than the front, likely because they are not subjected to as much natural wear as the front hooves.

Once the dirt is picked out, trim the overgrown part of the hoof almost flush with the sole of the foot, without coming to the quick. You will know you are close if one of your trimmings leaves

A goat's back hooves typically grow faster than their front hooves. This one is overdue for a trimming. *Cheryl Kimball*

Some breeds are polled (naturally hornless), while others grow horns. If you want a horned breed to be hornless, the kids should be disbudded, preferably before two weeks of age. These Oberhasli kids have recently been disbudded. *Kathleen Musser, Blue Ridge Dairy Goat Farm*

pink behind. You may find the toe of the hoof needs to be trimmed as well; you can do this last, first, or in the middle—just be sure to cut it straight across, getting rid of all the excess.

Horns

Some goat breeds are naturally polled, which means that members of that breed will never have horns. Other breeds are hit or miss: Some will grow horns, and some will not. Contrary to popular belief, horns are not exclusive to male goats; either sex can have horns.

Goats that have horns often spend a lot of time "sharpening" them. You will notice that they use most anything—the side of a doorway, a section of fencing, a tree—to run a horn up and down. And they will butt into things with their horns, like a plywood door or the side of the building. It seems that if the goat has horns, he or she feels the need to do something with them.

Hornless goats are desirable for the managed herd. Horned goats can be dangerous to other goats and to their handlers. A goat doesn't have to intentionally wield his horns to inflict harm; just tossing his head out of the way can inadvertently cause the horns to strike a bystander.

Also, a horned goat can get his head caught in fence openings. Even though he got his head in that way, he doesn't necessarily understand how to get his head out the same way. Many owners of horned goats tell stories of having to hacksaw a section of hog panel to release their goat: No matter how hard they tried to get the goat to turn his head to get it unstuck, the goat would have none of it.

There are a couple of ways to disbud kids so that their horns do not grow. Although a caustic powder was once commonly used, it is a slow and painful process. The most common method these days is to cauterize the horn bud with a hot iron. This is done before the kid is a couple of weeks old—old enough to have a visible horn bud area, but young enough that no horn has yet started to grow.

Disbudding is often the goat herder's least favorite task. It causes pain, and the kids holler like mad the whole time. Usually, however, within fifteen minutes of the procedure, they don't seem to remember anything happening.

There is an art to disbudding, and it is imperative that the kid stay completely still during the process. Someone very strong should restrain the kid, while someone with a steady hand and a good eye should operate the disbudding iron. You can also buy or make a restraining box that allows the kid's head to stick out.

Shave the hair away from the area around each horn bud. This makes it easier to see where you need to put the iron and you don't have to smell hair burning.

The iron tip needs to go completely over the horn bud and cauterize it completely. The iron is held on the horn bud for three to five seconds. You probably should plan to cauterize it two to three times, waiting a minute or two in between to let the iron reheat properly and give the kid a break. Then you need to do the other side. It is important to hold the iron on

These cute little horn buds will grow into big sturdy horns that can be annoying at best and weapons at worst. *Shutterstock*

long enough to cauterize the bud but not so long that you cause brain swelling. If you suspect brain swelling in a kid, administer an anti-inflammatory. Kids can die from swelling of the brain after disbudding.

The properly cauterized horn bud will scab over, and the scab will fall off on its own in a few weeks. If you do not feel up to the job, you can try to hire someone else to do it, but typically someone who is raising goats will simply learn how to do this unpleasant task. Goats without horns are a wonderful thing, and you will learn to appreciate the results.

If the disbudding did not get the entire horn bud, a "scur" will grow. A scur is a partial, often deformed, horn—perhaps the outer edge. Some scurs look like substantial horns, but they are not. Often they curl when they grow, and sometimes they curl around and head back into the animal's skull or even into the eye or ear, in which case they need to be cut. This is a messy and difficult job. The horn has a blood supply at the base and an inch or two up the horn, so trimming the scur can result in a lot of blood. This typically looks a lot worse than it actually is. Trimming a scur on a mature goat is best done by a veterinarian, with the animal under anesthesia.

Vital Statistics of the Goat

- Body temperature: 102.5 to 104 degrees Fahrenheit
- Pulse/heart rate: 60 to 80 beats per minute
- Respiration rate: 15 to 30 breaths per minute
- Puberty: 4 to 12 months
- Estrus ("heat") cycle: 18 to 23 days
- Length of each heat: 12 to 36 hours
- Gestation: 150 days

Shutterstock

Chapter 3

Goat Breeds

There are not as many breeds of common domestic goats as there are with other types of livestock. Of those described here, only the first eight are breeds recognized by the American Dairy Goat Association (ADGA). The ADGA breed standards are outlined at the beginning of the entry for those eight breeds.

The American Dairy Goat Association

The ADGA (www.adga.org), founded in 1904, is the breed-registering association for eight recognized dairy goat breeds. It keeps milk production records, sanctions goat shows, and provides breeding services such as DNA testing.

Dairy breeds are registered with the ADGA. *Daniel Johnson*

Opposite: The long-eared Nubian is one of the most popular breeds of goat in the United States. *Paulette Johnson*

Breeds Recognized by the ADGA

Alpine

Alpine doe. *Michelle Blair, Eagle's Wings Ranch*

Breed Standards:
Color: Eight different color patterns
Ears: Erect
Face: Straight (not Roman nosed)
Hair: Medium to short

There are three types of Alpine goat: French, American, and Swiss. The French Alpine was originally imported to the United States in 1922. Goats that display Alpine characteristics but have lineage that cannot be traced to these original French Alpine stock are known as American Alpine. The third Alpine type, the Swiss Alpine, has become known as the Oberhasli, which is now an ADGA-recognized breed of its own.

The Alpine goat is a medium-sized animal with a sturdy appearance, although its bone structure is more refined than other breeds such as the Nubian. Alpines average in height between thirty and forty inches (measured at the wither, or shoulder, area) and average in weight between 130 and 170 pounds. The higher end of the range is typically the males.

Alpines are hardy, easy keepers. They are also alert and smart and tend to be very people-friendly. The range of coat color is broad, but it does not include plain white or other color combinations that might indicate they are a mix of breeds.

The Alpine goat is a popular dairy breed. The butterfat content of their milk is low to moderate, at an average of around 3.5 percent.

The Alpine breed association is the Alpines International Club (www.alpinesinternationalclub.com).

Alpine kid. *Daniel Johnson*

Oberhasli

Formerly known as the Swiss Alpine, the Oberhasli is stockier in build than its American and French Alpine cousins. It also has a distinct coat color pattern of a reddish brown body (referred to as "chamois") with black points, as outlined in the ADGA breed standards below.

This breed has fought for survival, twice having lapsed in standards and assimilated back into the general Alpine breed. However, the breed association,

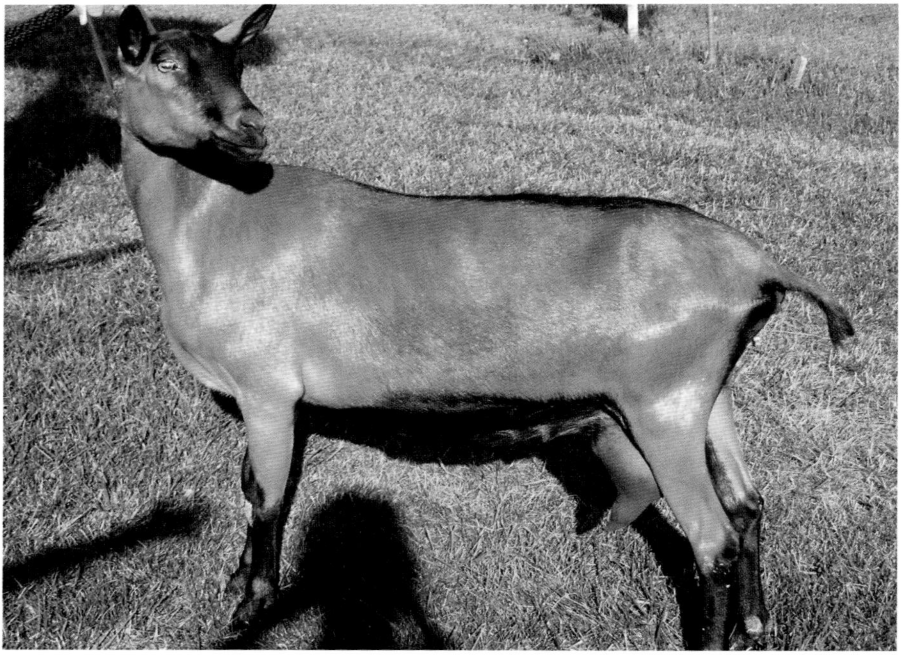

Oberhasli doe. *Natascha Jewell, Rainbow Basin Dairy Farm*

Breed Standards:

Color: Chamois preferred, although does may be black. A few white hairs through the coat and around the ears are permitted. Black stripes along face, dorsal stripe to tail, black belly, black below knees front and back, ears black inside.

Ears: Erect

Face: Straight or dished (not Roman nosed)

Hair: Short, thick haircoat

the Oberhasli Breeders of America (www.oberhasli.net), was finally established in 1976, and the ADGA gave the Oberhasli its own herdbook.

Oberhasli goats are typically raised for meat. However, their milk is good and said to be sweet in taste but low in butterfat content.

Oberhasli buck. *Debi Carroll, Ugotabkidn Goat Farm*

Nubian

Nubian. *Daniel Johnson*

The Nubian breed originated in the northeastern region of Africa known as Nubia. The goats were imported to England, where the breed was further developed; the resulting breed type became known as the Anglo Nubian. In the United States, it is known simply as the Nubian.

These handsome goats are popular with breeders. They are a versatile breed and can be used for meat, milk, and hide production. Although Nubian does typically have large udders, their milk production is somewhat lower than average; this factor is offset by a high butterfat content, averaging 5 percent. Another unique characteristic of the breed is that it tends to have a much longer

Breed Standards:
Color: Any color acceptable
Ears: Long, wide, pendulous
Face: Strongly convex (Roman) nose
Hair: Short, fine, glossy

Opposite: Nubian. *Paulette Johnson*

Nubian kids. *Paulette Johnson*

breeding season than other goats, making it possible, with good management, to have milk production year-round.

Nubians are a medium-build goat: The breed standard requires that does stand a minimum of thirty inches high and weigh on average 135 pounds, while bucks must reach a minimum of thirty-five inches and at least 175 pounds. They are popularly used under harness to pull carts. And something Nubian owners will not admit—but almost everyone at all familiar with goats knows—is that Nubians tend to love the sound of their own voices!

Saanen

Saanen kids. *Daniel Johnson*

A pure-white goat, the Saanen (pronounced "sah-nen") got its start in the Saane Valley of Switzerland. (See the Sable entry on pages 60–61 for a Saanen of a different color.) They are on the upper end of the size scale when it comes to goats, weighing in at an average 145 pounds and a thirty-inch minimum height standard for does. And their milk production follows suit: A Saanen doe averages over 2,000 pounds of milk in a typical dairy year (around 305 days) with a butterfat average of around 3.5 percent.

The National Saanen Breeders Association (www.nationalsaanenbreeders.com) is the very active breed association.

Breed Standards:
Color: White
Ears: Erect
Face: Straight or dished (not Roman nosed)
Hair: Short, fine

Saanen. *Daniel Johnson*

Opposite: Saanen buck. *Paulina Lenting-Smulders, iStockphoto*

Sable

Sable doe. *Ken and Janice Spaulding, Stony Knolls Farm*

Breed Standards:
Color: Any color except white
Ears: Erect
Face: Straight or dished (not Roman nosed)
Hair: Short, fine

Sable buck. *New Zealand Dairy Goat Breeders Association*

The Sable is basically a purebred Saanen that is not white. Sables are typically a creamy tan color but can be many colors. Sables are created genetically by breeding established Sable or by breeding two Saanens that possess a recessive-color gene. The Sable was recognized as a separate breed by the ADGA in 2005. For more information about raising this breed, go to the website of the International Sable Breeder's Association (www.sabledairygoats.com).

LaMancha

LaMancha doe. *Lorij Schmick, Sunshine Meadows Farm*

Breed Standards:
Color: Any color or combination of colors
Ears: One of two ear types:
- Gopher ear: maximum one inch long with little or no cartilage; this is the only type of ear allowed on a registered buck.
- Elf ear: maximum two inches in length, end may be turned up or down, some cartilage allowed.

Face: Straight (not Roman nosed)
Hair: Short, fine, glossy

LaMancha doe. *Lorij Schmick, Sunshine Meadows Farm*

The LaMancha breed in America originated from stock from Spain. The breed's official U.S. registry did not begin until the 1950s, even though recorded information about the "short-eared breed" in the United States had been in print for a while. In almost any discussion of LaMancha history in the United States, the name Eula F. Frey will arise. She is said to have originated the breed in the 1930s—which would make it the only "true" American goat—by crossing some small-eared goats with her purebred stock of Nubians and Swiss breeds. Whatever the real story, Frey was clearly a champion of LaManchas and the breed has thrived over the years.

The distinguishing characteristic of the LaMancha is its so-called lack of ears. The goat does, of course, have the anatomy for hearing; it is simply the exposed part of the ear, called the pinna, that is much diminished. The breed's ears come in two types: gopher and elf. The gopher pinna is less than one inch; the elf ear can be up to two inches long. Bucks can only be registered if they have gopher-type ears, as an elf-eared buck and an elf-eared doe can produce offspring with regular pinna.

LaMancha elf ear. *Jen Brown*

The LaMancha is known to be a friendly, easygoing, hardy goat and is popular in the show ring. The does are good milk producers, and their milk has a high butterfat content. They are of average size, similar to the Alpine.

The official LaMancha breed association is the American LaMancha Club (www.lamanchas.com).

LaMancha gopher ear. *Jen Brown*

Nigerian Dwarf

Nigerian Dwarf doe. *Michelle Blair, Eagle's Wings Ranch*

Breed Standards:
Color: Any color or combination of colors
Ears: Medium-length, erect
Face: Straight or dished (not Roman nosed)
Hair: Short, fine

This little goat hails from West Africa. The history of how the Nigerian Dwarf goat made it to the United States is, as with most goat breeds, sketchy at best. Many reports indicate they were first brought aboard ships as food for large animals, such as lions, on their way to American zoos. However they really got here, Nigerian Dwarves, like several other goat breeds, were fortunate enough to gain a few strong champions of the breed who ensured its development in the United States.

The diminutive size and friendly demeanor of Nigerian Dwarf goats make them great for children. They make manageable 4-H livestock projects. They are also good pets and surprisingly great milk producers.

Dwarf and pygmy goats have similar origins in West Africa, but there are major breed differences. The dwarf is bred to be proportional to its full-sized counterparts. While pygmies are often described as looking like "beer kegs with legs," Nigerian Dwarves simply look like

Nigerian Dwarf doe. *Cheryl Kimball*

small goats. They average a top weight of around seventy-five pounds. The registry standard for a dwarf doe is seventeen to nineteen inches in height and twenty-one to twenty-three inches for a buck.

Nigerian Dwarf goats can mostly cohabitate with a herd of full-sized goats, with minor additional fencing considerations to accommodate their ability to get through smaller spaces. A benefit to breeding dwarf goats for milk production is that they breed year-round, allowing a breeder to get three kiddings in two years. This gives the doe about a six-month break between pregnancies, while providing almost year-round milk

Nigerian Dwarf buck. *Michelle Blair, Eagle's Wings Ranch*

production. Dwarf does commonly begin their breeding life between eight months and one year of age; bucks can be used for breeding as young as three months and definitely well before they are a year old. Owners need to be aware, however, that these little guys are potentially fertile as young as seven weeks!

The Nigerian Dwarf also tends to produce multiple births. While twins are common in goats and triplets fairly common, dwarves might have four and even five kids at a birth. And these little dynamos can produce as much as three to four pounds of milk per day with a very high butterfat content of 6 to 10 percent.

Dwarf goats are registerable in three registries: the American Goat Society (AGS), the International Dairy Goat Registry (IDGR), and the Canadian Goat Society (CGS). Another organization for the breed is the American Nigerian Dwarf Dairy Association (www.andda.org).

Nigerian Dwarf kid. *Daniel Johnson*

Toggenburg

Toggenburg. *Daniel Johnson*

Breed Standards:
Color: Solid, varying shades from light fawn to dark chocolate
Ears: Erect and carried forward
Face: Dished or straight (not Roman nosed)
Hair: Short to long, soft, fine

The Toggenburg is credited with being the oldest registered breed of goat. Its herdbook was established in Switzerland in the 1600s. The breed arrived in the United States in 1893.

Toggenburgs have a distinctive coat color and markings. The coat ranges from a grayish taupe to other shades of brown. The markings include white stripes along each side of the face, white around the tail, and white inside the legs.

The Toggenburg tends to possess a nice temperament. It is an average-sized dairy goat with an above-average milk production and around 3.3 percent butterfat content.

Opposite: Toggenburg. *Daniel Johnson*

Other Breeds

The ADGA recognizes only dairy breeds, but there are several other goat breeds popular in the United States and Canada and kept for meat, fiber, or just as pets. Several of these are landrace breeds that have developed naturally over time and keep well on range, in fairly exposed conditions.

Pygmy

Pygmy. *Paulette Johnson*

Pygmy male. *Cheryl Kimball*

The Pygmy goat originated in West Africa. From there it was exported to Europe. The first documented importation of Pygmy goats into the United States was from Sweden to New York and California in 1959. The offspring of these goats ended up in zoos and medical research labs, as well as with individual owners. The Pygmy eventually spread throughout the United States and today is one of the most popular pet goat breeds.

Compared to a standard-sized goat, the Pygmy is not only smaller but atypically proportioned. It has a barrel-shaped body and short, bowed legs. An adult doe stands sixteen to twenty-two inches, and an adult buck sixteen to twenty-four inches.

These fun little goats are found in all colors, but the most common is the agouti pattern, which is a mix of dark and light hairs. While any color is acceptable to the breed, those that are not all black must have certain distinctive markings, such as eyes, muzzle, and forehead accented in a lighter tone than the body color.

Pygmy goats are good-natured and easy to care for. The breed association is the National Pygmy Goat Association (www.npga-pygmy.com).

Breed Standards:
Color: Any
Ears: Erect
Face: Dished (not Roman nosed)
Hair: Medium, straight, full coat

Pygmy doe and kids. *Paulette Johnson*

Pygmy kids and doe. *Paulette Johnson*

Well-Placed Teats

The phrase "well-placed teats" refers to teats situated on the udder in a good position for the animal to be milked and to be nursed by young. Teats that are not well placed—such as at odd angles or too close together to allow more than one kid to nurse at a time—add to the already time-consuming process of milking. They can also make it difficult for a kid to nurse, increasing the chance of the kid being undernourished.

Mary Schowe, iStockphoto

Boer

Boer doe. *Ernie and Roxanne Gray, Owner/Operator*

The Boer is the most popular breed of meat goat in the United States. It is indigenous to South Africa, where their registry formed in 1959. The Dutch colonized this part of Africa, and the name *Boer* is Dutch for "farmer." The goat was first brought to the United States relatively recently, in 1993.

Breed Standards:
Color: Reddish-brown head, white body
Ears: Medium-length, drooping
Face: Convex (Roman) nose
Hair: Short

Boer buck. *Copper Creek Boers, Oregon*

The Boer is a sturdy, stocky breed, bred for fast growth and heavy muscling for the meat market. A doe weighs as much as 225 pounds and a buck 300 pounds. Like the Nigerian Dwarf, the Boer doe has an extended breeding season, making it possible for a doe to have three kiddings in two years.

Founded in 1993, the official breed association is the American Boer Goat Association (www.abga.org).

Boer kid. *Cheryl Kimball*

Savannah

Savannah doe. *Tim Swain, 2T's Farm*

The Savannah originated in South Africa alongside the Boer goat. It has an all-white hair coat over black skin, which protects it from the sun. A landrace goat bred for meat production, the Savannah is extremely hardy, fast-growing, and muscular. Like other meat goats, it is typically raised on range. The Savannah has been bred in Africa since the 1950s but was first imported to the United States in the 1990s. Its breed society is the North American Savannah Association (www.nasavannahgoatassn.com).

Breed Standards:
Color: White preferred, red or roan allowed
Ears: Medium-length, drooping
Face: Convex (Roman) nose
Hair: Short

Spanish

Spanish buck. *Gurney Davis*

Breed Standards:
Color: Any
Ears: Long, drooped, close to head
Face: Straight or slightly convex
Hair: Most short-haired, some producing commercial-quality cashmere

Spanish doe and kids. *Gurney Davis*

The Spanish goat is a hardy breed that does well in hot climates and rugged terrain. Brought to the Americas in the 1500s by Spaniards, they are relatively disease- and parasite-resistant and take good care of themselves.

The Spanish goat was popular in the Southwest and the Southeast in the 1800s. When cows, pigs, and chickens took over the livestock industry in the mid-twentieth century, these goats began to disappear. There has been an attempt to re-establish the breed, and in 2007, the Spanish Goat Association was established (www.spanishgoats.org).

Myotonic

Myotonic doe. *Laurie Macrae, White Sage Farms*

The so-called fainting goat, Myotonic goats are also known as stiff leg, wooden leg, or Tennessee fainting goats. Breeders have selectively bred them into two strains: a smaller animal found mostly in Tennessee and the eastern United States and a larger animal bred for the meat market and found mostly in Texas. Breeders are attracted to the Myotonic goat because of its unique characteristic of "fainting" when frightened or excited—

Breed Standards:
Color: Any
Ears: Large, horizontal but not drooping
Face: Dished
Hair: Mostly short-haired, but can be long-haired and even produce commercial-quality cashmere

Myotonic buck and kid. *Laurie Macrae, White Sage Farms*

it literally falls over and lies stiff for a few seconds. Myotonia is defined as a spasm or temporary rigidity of the muscles. Physiologically, the chemicals normally released in the automatic fight-or-flight reaction do not get released in the Myotonic goat.

This "fainting gene" is recessive, so when Myotonic goats are crossbred with other breeds, the resulting offspring does not express this trait. The American Livestock Breed Conservancy has added the Myotonic goat to its conservation priority list of rare breeds.

The International Myotonic Goat Herdbook is kept by Pedigree International (www.pedigreeinternational.com).

Myotonic doeling in faint. *Laurie Macrae, White Sage Farms*

Myotonic kids. *Laurie Macrae, White Sage Farms*

Rare Breeds

There are many breeds of livestock that are considered endangered, threatened, or rare. The American Livestock Breeds Conservancy (ALBC) (www.albc-usa.org), founded in 1977, is an organization dedicated to conserving rare breeds with an eye toward genetic diversity and biodiversity. The ALBC identifies seven goat breeds on its Conservation Priority List:

- **Critical**: Arapawa, San Clemente
- **Watch**: Myotonic, Spanish
- **Recovering**: Nigerian Dwarf, Oberhasli
- **Study**: Golden Guernsey

Kiko

Kiko buck. *Mia Nelson, Lookout Point Ranch, Oregon*

The Kiko goat originated as a landrace breed in New Zealand. In the 1980s, farmers crossed it with domestic dairy goats to develop a meat breed that has exceptional milking capability. The domestic Kiko was first imported to the United States in the 1990s.

Kikos are hardy animals typically raised on range with no need for supplementary grain to bring to market weight. Bucks have long horns with a distinctive spiral; does have horns that sweep back. The breed is primarily identified by its performance characteristics: rapid weight gain, heavy muscling, and parasite resistance. The offspring of a Kiko and a Boer is called a Genemaster.

The breed societies are the Amercan Kiko Goat Association (www.kikogoats.com) and the International Kiko Goat Association (www.theikga.org).

Breed Standards:
Color: White common, any color allowed
Ears: Medium, set high
Face: Straight (not dished or Roman nosed)
Hair: Short to long

Arapawa

Arapawa buck. *Cheryl Kimball*

The Arapawa goat is a landrace breed with a complex history that has brought it to near extinction. Unlike most domestic breeds, which are created through controlled matings, landrace breeds form naturally in the wild in response to environmental factors. Once populating Arapawa Island off New Zealand, Arapawa goats were considered nuisances that decimated the island's forests. In the 1970s, attempts were begun to eradicate them through hunting. An island resident, Betty Rowe, began to try to save

Breed Standards:
Color: Brown-and-black patchwork
Ears: Erect
Face: Straight
Hair: Short, fluffy

Arapawa doe. *Lori A. Corriveau*

them. As a result of her hard work, a few now remain on the island in a sanctuary. The rest have been collected by Arapawa enthusiasts around the world, including three bucks and three does finding their way to Plimoth Plantation in Massachusetts in 1993.

There are only a few hundred Arapawa goats in domestication worldwide. With such a small population, it is difficult to breed these goats without the complication of inbreeding. However, devoted Arapawa owners are making sure these distinctively colored brown-and-black patchwork goats make a full comeback. Their small, sturdy stature and hardy nature are considered important traits to preserve in the overall goat genetic pool.

For more information on this rare breed, consult the International Arapawa Goat Association (www.arapawagoats.com).

Arapawa kid. *Lori A. Corriveau*

San Clemente

The San Clemente goat is a feral breed that has lived for centuries on San Clemente Island, off the coast of California. The island used to be owned by the U.S. government and has been managed by the U.S. Navy since 1934.

Like Arapawa goats, San Clemente goats overran the island and were considered a nuisance and a threat to the indigenous plant species. In the 1970s, the navy began systemically removing the goats from the island by forcible methods, including shooting them from helicopters. Animal-rights groups rallied to remove the goats from the island without killing them; the last goat left the island in 1991.

Breed Standards:
Color: Reddish brown or tan with black markings
Ears: Erect
Face: Straight
Hair: Relatively long-haired

These relatively small goats are used predominantly as a meat animal. They are a rare breed, with only a few hundred distributed throughout the United States and Canada. Their breed organization is the San Clemente Island Goat Association (www.scigoats.org).

San Clemente goats. *Mehgan Murphy, Smithsonian's National Zoo*

Cashmere

The Cashmere goat is not a breed of goat at all. Any goat's undercoat is "cashmere" fiber. The outer coat is known as guard hair. The so-called Cashmere goat is one whose undercoat produces commercial quantities of cashmere. Although any breed can produce cashmere, a commercially viable Cashmere goat herd requires careful breeding and meticulous culling to ensure high-quality fleeces.

Breed Standards:
Color: Any
Ears: Any
Face: Any
Hair: Fine, downy undercoat of commercial quality

Cashmere goat. *Baldur Tryggvason, iStockphoto*

Cashmere goat. *Baldur Tryggvason, iStockphoto*

Cashmere is a valuable fiber, and the industry is carefully regulated and complex. The U.S. Wool Products Labeling Act of 1939 (U.S.C. 15 Section 68b(a)(6)) specifically defines cashmere:

1. (A) the fine (dehaired) undercoat fibers produced by a cashmere goat (*Capra hircus laniger*);
2. (B) the average diameter of the fiber of such wool product not exceeding 19 microns; and
3. (C) contains no more than 3 percent (by weight) of cashmere fibers with average diameters that exceed 30 microns. The average fiber diameter may be subject to a coefficient of variation around the mean that shall not exceed 24 percent.

Quite a specification!

But anyone who decides to raise Cashmere goats will have just as much fun and frustration as raising any other goat. They do require shearing once a year, in the spring, just before they would shed their long coat. That coat serves them well in even the coldest winter, making them hardy animals that need minimal housing.

The U.S. cashmere industry is a growing niche market. Several organizations have formed to help grow the industry, including the Eastern Cashmere Association (www.easterncashmereassociation.org), the NorthWest Cashmere Association (www.northwestcashmere.blogspot.com), and the Texas Cashmere Association (www.texascashmere.com), as well as the International Cashmere and Camel Hair Manufacturer's Institute (www.cashmere.org).

Angora

The white Angora goat originated in southwest Asia and came to the United States in the mid-1800s. A colored variety of the Angora was imported in the late 1900s; it was raised mainly in the Southwest and came to be known as the Navajo Angora.

Breed Standards:
Color: White, black, or faded red
Ears: Drooping
Face: Dished (not Roman nosed)
Hair: Long, silky mohair

White Angora. *Daniel Johnson*

Black Angora buck. *Daniel Johnson*

A fiber breed, the Angora grows a hair coat called mohair, unlike all other goats, which grow cashmere hair. Commercial-quality mohair is sold for hand-knitting and used in specialty fabrics. Although Angoras are sometimes mistaken for sheep, mohair fiber is much longer and silkier, not short and kinky like sheep fleece.

Fiber goats are typically raised on range. When raised in a cold climate, they require protection from the elements in the month following shearing. Goats are sheared twice a year.

The traditional Angora goat has a white hair coat and light-colored horns. Its registry is maintained by the venerable Angora Goat Breeders Association (www.aagba.org), founded in 1900. The Navajo Angora Goat Record (www.navajoangorgoat.org) promotes this rare variety. Farmers now also breed colored Angoras in shades of black and faded red. The Colored Angora Goat Breeders Association (www.cagba.org) was founded in 1998.

Opposite: Faded red Angora kid. *Daniel Johnson*

Nigora

Nigora buck. *Betty Moon, Mystick Acres Farm and Rabbitry*

Breed Standards:
Color: All colors/markings
Ears: Medium, erect or drooping
Face: Straight or slightly dished (not Roman nosed)
Hair: Three types of fleece of commercial quality acceptable:
- **Type A**: Mohair (long, lustrous, fine, silky, smooth)
- **Type B**: Mohair/cashmere blend (medium, curly, lustrous, soft, airy)
- **Type C**: Cashmere (short, fine, matte)

Nigora doe. *Betty Moon, Mystick Acres Farm and Rabbitry*

The Nigora goat was developed from mohair-producing Angoras crossed with cashmere-producing Nigerian Dwarves. It is currently being developed as a dual-purpose breed, used for both fiber and milking ability. The Nigora may be horned, polled, or disbudded.

The Nigora's fiber can come in three types, depending on which ancestor it favors. It can be more mohair (Type A), more cashmere (Type C), or a fluffy blend of the two (Type B).

Established in 2006, the American Nigora Goat Breeders Association (www.nigoragoats.homestead.com) maintains an open registry that allows breeders to grade up. This means that in addition to Angoras, other fiber goats (Pygora, Pygora-Colored Angora, Cashgora) may be crossed to Nigerian Dwarves in a Nigora breeding program. Preference is given to registered purebred stock. ADGA dairy breeds, meat breeds (Boer, Kiko, etc.), and myotonic goats may not be used.

Pygora

The Pygora goat began as a cross between a Pygmy and an Angora goat, bred by Katherine Jorgenson in Oregon, who eventually formed a registry. She was looking for a fine hand-spinning fiber. The coat is described in four color patterns: black, white, caramel, and agouti (black-and-white mixed hairs). Disbudding these naturally horned goats is acceptable.

The offspring of an Angora and a Pygmy is not a true Pygora but just a first-generation (F1) cross. The offspring of two F1s is considered a Pygora. Genetically, a Pygora is typically 50 percent Angora and 50 percent Pygmy, but it can be up to 75 percent of one breed or the other.

They are hardy animals and are considered easy to handle and friendly. Check out the Pygora Breeders Association (www.pygoragoats.org) for more information.

Pygora doe. *Fran Bishop, Rainbow Spring Acres, Pygora Breeders Association*

Pygora buck. *Fran Bishop, Rainbow Spring Acres, Pygora Breeders Association*

Breed Standards:
Color: Four color patterns acceptable (black, white, caramel, agouti)
Ears: Erect or drooping
Face: Dished (not Roman nosed)
Hair: Three types of fleece of commercial quality acceptable:

- **Type A**: Mohair (long, lustrous, fine, silky, smooth)
- **Type B**: Mohair/cashmere blend (medium, curly, lustrous, soft, airy)
- **Type C**: Cashmere (short, fine, matte)

Breeds Found in Other Countries

Some three hundred domestic goat breeds exist worldwide, only a handful of which are found in North America. The rest are scattered throughout the world. These are some of the more well-known international breeds:

The **Bagot** is a striking goat named for the Bagot family of Straffordshire, England. It is a medium-sized goat with a long hair coat and horns, a straight face, and erect ears. The breed is known to be skittish. It has a beautiful color pattern—all black from the nose to the shoulder area, and white from there back. There are sometimes patches of black in the rear quarters, but these are considered undesirable. Bagots do not have a distinct agricultural purpose, but they could be used as meat goats. The Bagot is much reduced in numbers and considered critically endangered by the British Rare Breeds Survival Trust. The breed organization is the Bagot Goat Society (www.bagotgoats.co.uk).

The so-called **Australian** goat is the domesticated version of the feral goats living in the wild in Australia. The population got a bit out of hand when the fiber industry collapsed in the 1920s, and Cashmere and Angora herds were set free. Most of the feral Australian goats are good meat and fiber producers. They have adapted well to their environment and have become sturdy, healthy animals. They are nicely crossbred with domestic stock. Besides being used for meat, fiber, and leather, they are also being used for weed and brush control.

The **Dutch Landrace** goat originated as a feral breed in the Netherlands. By 1958, it had been reduced to merely two animals. Through careful management, these two animals were used to re-establish the breed, bringing it back to a population of more than one thousand animals by 1999. A Dutch Landrace herdbook was established in 1972. The Dutch Landrace is multicolored and medium in size, with horns, erect ears, a straight face, and long hair. They are mostly used for vegetation management, since they fare well in adverse conditions. They are typically compared against the Toggenburg, the standard breed of the Netherlands. Any color is acceptable, except Toggenburg markings. The Dutch Landrace has much lower milk production than the Toggenburg but a milk higher in butterfat.

The **Golden Guernsey** goat is a dairy breed whose milk's high butterfat content is perfect for cheese-making. This rare goat has an uncertain history, but as is often the case, one breeder was largely responsible for its survival. Miriam Melbourne first saw the goats on the Channel Islands off the coast of Great Britain. She began raising them in 1937 and developed a breeding program for the Golden Guernsey in the 1950s. Nonetheless, the breed continues to be a member of the Rare Breeds Survival Trust of Great Britain. The Golden Guernsey is considered the perfect backyard or household goat, being small in stature, friendly, and producing a sufficient amount of milk for the average family. These goats are a beautiful golden color with occasional white markings. They have straight, erect ears and are usually naturally hornless. The official breed organization is the Golden Guernsey Society (www.goldenguernseygoat.org.uk).

The **Jining Grey** of Shandong Province, China, is often raised for the rare wavy pelt it has as a kid. The haircoat is black, white, or a combination of the two. They are prolific breeders, reaching sexual maturity at a very young age and commonly producing

Phyllis Clayton, Golden Guernsey Goat Society

three kiddings in two years. They are small in size, with does averaging less than sixty pounds and bucks around seventy-five pounds. They also produce some commercial-quality cashmere and are sometimes kept for this purpose.

The **Jamnapari** goat of India is reducing in numbers, with no more than five thousand purebred animals believed to be in existence. Named after a river in India, the Jamnapari is related to the Nubian goat and has long, drooping ears. It is a large, horned goat bred for both milk and meat. Both sexes have long skirting hair along their hind legs. The coat is white with tan patches. Although originating in India and Pakistan, the Jamnapari is now being imported into Indonesia and Malaysia with success. They have a high rate of conception, with triplets and quadruplets common.

The **White Shorthaired** goat originated in the Czech Republic. It developed in the early 1900s, when breeders began improving a landrace breed by crossing it with a Saanen buck. The White Shorthaired goat became recognized as a separate breed in the mid-1950s. The breed standard requires a pure-white coat with no other coloring. A dairy goat, it is average-sized and naturally polled for the most part. The population numbered over one million after World War II, but it is now thought to be around 35,000.

Chapter 4

Goat Behavior

Goat behavior is not heavily studied the way other livestock behavior is. Horses have several study groups, including the Equine Research Foundation in California. Cattle have been studied by people such as Temple Grandin in an effort to make their experience in feedlots and at the slaughterhouse less stressful; her interesting books include *Thinking in Pictures* and *Animals in Translation*. But ask any goat owner, and they will agree that goat behavior could perhaps be the most interesting of livestock behavior to study!

Curiosity

Goats are especially curious animals. Put something new into the goat yard and every goat will have to check it out. If you place something new outside their pen, be prepared for the security of your fences to be thoroughly tested.

While goats are definitely concerned about anything that seems threatening, they are curious enough that they can quickly get over that concern once they are pretty sure there is no threat.

Goats make great animals for petting zoos because this high curiosity level makes them curious about new people and very friendly.

Fear

When a goat feels threatened, it will—like its wild counterpart, the deer—stomp one front foot and snort. Another interesting thing about goats' response to an intruder

Prey animals in the wild, domestic goats remain alert to predators on the farm. *Shutterstock*

Opposite: Goats are so curious that they can be very difficult to photograph! *Shutterstock*

is that, unlike sheep that will flock together and derive safety (real or not) in numbers, goats will separate. This makes them impossible to "herd."

Kids often freeze when threatened by, for instance, a predator. This instinctive behavior may make the kid less visible to a predator. Young animals often have no scent, which prevents them from being detected by predators that rely on scent detection.

Shelter

One important thing for any new goat owner to know is that goats melt in the rain.

Of course they don't; but you would think they do if you ever watched a goat skedaddle for cover at the first couple of raindrops. A goat caught in rain gets a disgusted expression on its face, holds its ears back a bit, and tiptoes like the rain will melt its feet on contact with the ground as it makes its way to shelter. Goats actually don't seem to mind light snowfall, but rain is usually just intolerable.

Catching and Leading

Catching goats can be a problem, unless you have very tempting food to offer. In that case, you should not have a problem catching a goat at any time.

Leading can be a different story. Although food can certainly help, you will soon realize how strong goats are. If a goat does not want to go in the direction you want to lead it, you will be going in the direction it wants to go!

The best way to teach goats to follow a lead willingly is to start when they are very young kids. They are not as strong at that age (although even kids can be surprisingly strong), and they are more curious and eager and tend to love to be with people. So put a halter on them and strap a leadrope to it as soon as possible. It's best to use a rope halter around the head instead of a collar, since goats can very easily be choked.

Trick Training

Goats are quite capable of learning the same type of tricks that are associated with dogs. Goats are so food motivated that they will do most anything for a handout. You can teach a goat to kiss, shake hands, stand on its hind legs, or do tricks with props such as hoops, stools, and balls.

A goat kid is the definition of capricious! *Shutterstock*

Have fun with your goat and make use of this playful behavioral trait!

Goats have also been known to be housebroken, although you might check with all members of the family before testing this theory.

Kids

Watching young goats at play is addictive. These carefree little bundles of energy kick up their heels and define the word "capricious" (which, according to Webster, means "given to changes of interest or attitude according to whims or passing fancies"). Kids and adult goats love to climb on things and should have plenty of "mountains" in their pen for that purpose: large rocks, picnic tables, plastic children's furniture. Anything sturdy and secure will do. If there are slow-moving sheep available, kid goats will climb on them!

Like most young animals, kids—especially bucklings—play their reindeer games. Butting heads, which is almost always started with a good windup and rearing up on hind legs, is a good goat game that prepares young males to defend their does and their food territory—which of course they won't need to do in your goat yard; but hey, one never knows.

Kids play-mate and start at a very young age, even as young as a few days. They don't know what they are doing, and they do it to everything, male or female, but they will do it.

Competition for food starts almost immediately after birth, if there is a set of twins or triplets. They will each compete for the best teat and the best position under the dam.

Food

Goats love food. And they absolutely LOVE the foods that they love! The foods they love aren't the same for all goats, but you will quickly learn what those items are for individual goats. Some goats love popcorn, some love corn husks, some love stripping the post–December 25th Christmas tree (although be certain the tree has not been treated with preservatives).

Goats will climb on whatever is available. *Shutterstock*

Kids practice butting heads and learn to protect their territory as adults. *Shutterstock*

Chapter 2 covers goat care and setting up a pen. It is important, particularly with a large herd, to consider common goat behavior and everyone's safety during feeding time. For example, keyhole-type hay-feeding stations that have large keyholes can be dangerous; if a goat puts her head through the keyhole to eat and another goat comes along and tries to use the same keyhole, the first goat will get trapped on the bottom. She can get seriously hurt or even killed, either by being trampled or strangled in an attempt to get out. Be aware of these possibilities with animals like goats that are very food-oriented; awareness can help avoid a lot of grief.

If your goat herd is large, you may need to divide the animals up by their food behavior. Free-choice hay is usually not a problem, but grain usually instigates bad behavior! A herd of milking does typically get their grain individually in the milk room, so competition is not an issue. If you have any wethers in, say, a brush-grazing herd, feeding them grain with ammonium chloride additives can help alleviate urinary calculi problems (see chapter 2)—and this grain-feeding time can get testy. Bucks also need grain, especially around the breeding season when they are using up a lot of energy on mating.

Dominance

The main factor with feeding time problems is dominance. Who is dominant over whom, commonly called (from poultry) a "pecking order," is a game that is always being played out among most livestock of any kind. As animals age and new animals come into the herd, each member's level in the pecking order is constantly being challenged and changed. Although goats rarely get into actual fights, they can accidentally hurt each other as they attempt to claim their rung on the herd ladder.

Mixing very young animals with middle-aged ones can cause problems. The older animals can accidentally hurt the younger animals as they assert their dominance, simply due to sheer size difference. Putting young animals in with much older animals usually doesn't cause this problem, as the older animals are typically secure in their rank and don't want to expend all the energy it takes to knock around other goats. However, the opposite can happen: The young goats can be too much for the older goats, climbing on them and trying to engage them in caprine games. Keeping an eye on the herd dynamic is important.

Aggression

Goats are rarely mean or aggressive to humans, and those that are have often been made that way by being teased, roughhoused with, or mistreated. If a goat

Goats love food and will search far and wide—and apparently high—for anything that comes close to being edible. *Shutterstock*

does threaten you with butting, take this behavior very seriously and discourage it immediately; being butted hard by a goat can break a bone. Numerous methods are promoted for discouraging butting. Some goats will respond to simple methods like "answering back" with dominant language, or even mimicking butting. You can also try deterring the unwanted behavior by using a spray bottle filled with water (thus taking advantage of goats' aversion to water).

To make sure the goat understands that you are the dominant party, some suggest laying the aggressive goat down and holding it there until it stops struggling and becomes submissive. You can accomplish this by standing on one side of the animal, reaching through its legs, and grabbing the two legs on the other side, pulling them toward you, much like you would to set up a sheep for shearing. Make sure the goat has a soft place to land. Stand with your foot or knee on the goat's neck, gently enough not to hurt it but strongly enough that a struggling goat can't get up. When the goat is calm and allows you to pat it, you can let it up. That may be the end of your problem, it may require another round, or it may never work, but if you have a serious butting problem you will need to try things until you figure it out.

Another last resort is to borrow a dog shock collar with a remote control, and use it when the goat exhibits the butting behavior. You probably won't have to reinforce this more than once or twice before solving the problem.

Introducing New Animals to the Herd

In addition to the health concerns mentioned in chapter 2, there are also behavioral concerns to consider when introducing new animals to an existing herd. As mentioned earlier, it is often good to keep animals of similar ages together. And keeping them penned by sex can be useful, although it depends on the temperament of the individual animals.

Bucks are generally kept separate from other parts of the herd, not only because of their behavior toward other goats, but also because of their personal habits, which are explained in the next section. The exception is during breeding season, when the buck is put in with the does.

Wethers are typically docile animals, but they can exhibit the same dominance behavior as other goats, especially about food. And since they often grow to be larger in size than does, an aggressive or assertive wether may need to be separated from the herd or culled.

Kids love to hand-feed goats, but beware of the animals becoming aggressive about getting to the food, especially when there is more than one goat present. *Shutterstock*

Goats are basically friendly, curious animals (*above*). But they have a strict pecking order. Pay close attention to newly introduced goats in your herd, and separate any goats that are harassing or being harassed (*below*). *Shutterstock*

Buck Hygiene

Bucks are kept in separate pens away from the does for obvious reasons: to prevent out-of-season pregnancies. But they also have personal hygiene habits that make them unwelcome in the barnyard. An intact male goat likes to scent himself by urinating on himself, including his front legs and beard, making him extremely odiferous. This apparently is the Old Spice of the goat world—the does love it. Or so the bucks think. But humans typically don't love it. So plan to have a buck pen some distance from the general barnyard and definitely away from your house—and away from your neighbors' houses as well!

Bucks can often be kept together and get along just fine, which is nice for them and nice for you, not having several stinky places on the property.

Does in Heat

You usually don't have to guess when a doe is in heat. She tends to be very vocal about it, yelling her needs and desires to anyone within hearing range of a loud goat. A doe in heat also "flags" her tail a lot, wagging it rapidly and often.

Some does also have other habits such as being aggressive or "affectionate" with other herd members. Some don't eat as well when they are in heat. The heat cycle only lasts a few days, so whatever annoying behavior an individual exhibits will be over soon—but they will come back again in about three weeks.

You will quickly learn the heat behavior traits of your does. If you have a milking/breeding herd, these behaviors help you better target your does' heat cycles. If you do not have or do not plan to have a milking herd or need offspring for meat, then heat behavior can be a good reason to keep a herd of wethers instead!

Mating Behaviors

The goat is a symbol of lechery for a good reason. Bucks are obsessed with sexual activity, especially during the fall breeding season. Goatworld.com describes the sexual behavior of the buck very well:

> *He will often be sniffing the urine of does, extend the head and neck into the air with the upper lip curled up (the "Flehmen" posture), searching for the olfactory and gustatory stimuli that indicate to him that the doe is in estrus.*
>
> *Upon identifying a doe in estrus, the buck will follow her, and then move up in an attempt to herd the doe away from the rest of the flock. Once separated the buck will begin to paw the ground around the doe in an apparent display of masculinity. During these and subsequent stages of precopulatory behavior, the buck emits a frequent hoarse "'baaing" that is often termed a "grumble". The buck can also be observed to run his tongue in and out of his mouth during these first two stages and is generally very excited. Next he proceeds to sniff and nuzzle the genital areas of the doe, while intermittently rubbing against the side of the doe.*

For her part, a doe in heat will rub against the buck, apparently enjoying his urine cologne. Copulation is extremely quick. If the doe is receptive, you could miss the breeding in the proverbial blink of an eye. On the other hand, the inexperienced or not-quite-interested-enough doe will not stand still and will start to walk around while the buck is mounted and attempting to thrust.

A stinky buck curls his lip, trying to identify a doe in heat. This is called the Flehmen posture.
Shutterstock

Milking

Observing goat behavior in the milking parlor is an intriguing activity. Outside of the agricultural world, a lactating doe's udder would be relieved throughout the day from the nursing activity of the offspring that instigated the milk production to begin with. However, in the dairy environment, the kids are removed within a day or two of birth.

At that point, udder relief is taken over by the hand-milker or the milk machine and is done only twice a day. By about the tenth or eleventh hour, the doe is getting pretty uncomfortable. Go to any dairy—cow or goat—within an hour or two of the normal milking time, but especially when the animals hear activity in the milk parlor, and you will see the milking animals all lined up eagerly at the parlor door.

This routine is established because the goats have learned that every twelve hours they get relief from this engorged udder they are carrying around. But the milking parlor produces other routine behaviors as well.

That line of goats at the door to the milk parlor is almost always in the same order each milking. The milking pecking order is established, and does tend to file into the same position in the milk stand each day.

Prior to milking, some farmers will "settle" the herd with a hay feeding of particularly luscious hay. The animals get to know that routine as well and line up at the hay feeding area.

Routine

In general, goats love routine and will pick up on a schedule of activities

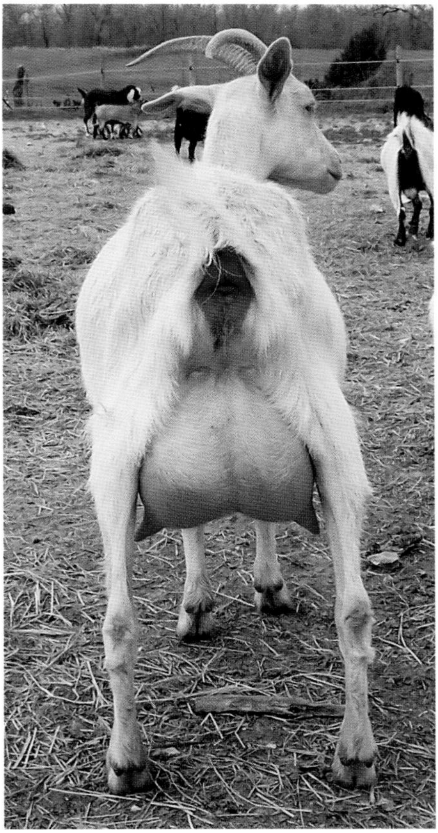

The full udder of a milking doe makes her eager to line up for the barn at milking time. *Michelle Blair, Eagle's Wings Ranch*

very quickly. In fact, if you start to do something regularly, be prepared to be tormented by very loud goats if you one day don't continue the routine. This routine behavior can include something like feeding a treat midafternoon every day when you go out to get the mail. Either plan to do this on Sunday, too, or the cacophony of displeased goats, not understanding that you don't need to get the mail on Sunday, will disturb your peaceful neighborhood. Or the behavior can be letting your two pet goats out every morning around ten to browse until noon. At 9:45 you will have eager goats waiting at the gate; at 10:10 you will have noisy ones! So don't start a goat routine you don't want to continue—or you will have to tolerate their expression of displeasure until they forget, which will take a while.

Illness

An ill goat behaves differently from one that is healthy. It is said that by the time you notice the symptoms, the goat is probably very sick. A goat's natural instinct is to hide its illness from the world, as a sick goat in the wild is a prime target for a predator. But if you are familiar with the normal behavior of the animals in your herd, you will notice a sick goat pretty quickly.

Some clear signs of illness include the following:

- **Disinterest in food.** Being the food-oriented animals that they are, goats are seldom disinterested in food. And goats in a herd are often in constant competition regarding the feed supply, so it's even more unusual for them to be uninterested in mealtime.
- **Runny or other unusual stool.** Normal goat manure is pelleted. Anything other than that is a sign that something is not right. If you have a large herd, you may see strangely shaped manure in the goat pen but not know right off which goat is producing it. Look for signs of diarrhea on someone's rump. Stay in the pen for a while and try to observe each herd member defecating. (Obviously this will be impossible with a huge herd in large acreage, but if you

have a smaller herd or a herd that is divided by age or sex groups, it is a little easier.) You shouldn't have to wait too long; goats defecate pretty regularly!

- **Hanging out in the corner away from the herd**. This is a sign of something wrong, especially if the goat is normally a vibrant member of the group.
- **A change in posture**. This sign may be less obvious than others, but it is something that an experienced goat person—and everyone who gets into goats will become an experienced goat person eventually—will notice readily. Look for things like a drooped head or a hunched back.
- **Getting up and lying down frequently**. This behavior is a sign of pain frequently exhibited by males who are having urinary blockages from calculi buildup.
- **Unusual vocalization**. A strange call is a clear sign that something is wrong.

Call for help immediately if you observe these signs in your goat herd. Goats can go downhill fast, and unless you are so experienced you know what clear action to take, you will be putting your goat's life in jeopardy if you do not get immediate help.

Opposite: Keep your goats happy and healthy. *Shutterstock*

Chapter 5

What Can You Do With a Goat?

People raise goats for myriad reasons. And often those first couple of pet goats obtained for some obscure purpose become the beginning of an addiction. However, there are also plenty of legitimate uses for goats.

Milk

We are in a new age of a penchant toward locally grown, sustainable agriculture. We want to know where our food comes from, how it is produced, and what kind of chemicals are used (or not) in the production of our food—we just plain want to become more in touch with the food chain (that is, the meat we buy in the plastic wrap at the grocery store did not magically appear there).

With that in mind, people who never thought of themselves as "farmers" of any kind are considering getting one or two animals to grow their own food—a pig or cow for meat, a few chickens for some eggs, and perhaps a goat for milk.

Getting one goat for the family dairy needs has long been precipitated when a child comes into the family who is lactose intolerant. Cow milk is high in lactose, a type of sugar that is less water soluble than other sugars, such as glucose. Although lactose *is* present in goat milk, goat milk overall has a much lower lactose percentage and is more easily tolerated by those who have trouble digesting lactose.

Depending on the breed chosen, the individual goat, and your nutritional program, the family milk goat can provide up to several quarts of milk a day.

In the dairy environment, kids are typically removed from the dam after a day or two and bottle-fed several times a day. *Shutterstock*

Opposite: Angora goats are raised for mohair fiber. *Daniel Johnson*

Dairy

The most common agricultural pursuit with goats is to establish a goat dairy. Goat dairies have become more and more popular. Goats take up less space than cows, they eat less, their milk is less common and therefore more "gourmet" (that is, it can sell at a higher price than cow milk!), and it's low in lactose, which means goat milk is great for the growing number of children and adults who are lactose intolerant.

Goat milk can be sold to consume as milk, but there are a few lucrative side products made from it, all of which we'll get to in a moment. But first, there are some important things to know before setting up a goat dairy.

Goat milk and goat cheese, called chèvre, are products of the dairy goat. *Shutterstock*

In the Raw

Selling raw (unpasteurized) milk is legal in about half of the fifty U.S. states. Be sure to check the regulations in your state before deciding to sell raw milk from your farm.

Shutterstock

The Facility

Producing milk is not something you can do cavalierly. To have a working goat dairy, you need to think ahead about an organized setup. Cleanliness reigns supreme. Milking animals are routine-oriented, and your facility needs to accommodate routine.

You will need a milking parlor where the animals come in to be milked. There should be as many milking stations as the number of milkers (of the human variety) you plan to have—a small dairy will probably have just a couple of milking stations, either hand or electric.

Your facility must have appropriate storage for milk. Milk needs to be cooled down almost immediately to stop bacteria production and reduce the chance for milk to become tainted by barn smells.

If you are generating milk in any significant quantity, you will need a storage tank. And if your milk is being sold for consumption as milk, you will need to arrange to have it picked up by a milk processor. The Goat Dairy Library suggests a four-day pickup schedule. In order to accommodate that length of turnaround, you need to have your cold storage at a temperature of 33 degrees Fahrenheit (1 degree Celsius) to prevent bacteria growth.

Dairy equipment must be washed and sterilized after each use. This process is now automated in most dairies; anything the milk touches is never exposed to outside air. *Cheryl Kimball*

Via Lactea milks at 6 a.m. and 6 p.m. The milking does begin to line up at their pasture gate around 5:30. *Cheryl Kimball*

Dairy Equipment

There are several important pieces of equipment you will need for a goat dairy, including the following:

- Milk stands
- Milking machine(s), unless you plan to do it all by hand
- Refrigerated milk storage

Dairy Supplies

Some smaller supplies are also important. These include the following:

- Wipes for cleaning udders before milking
- Teat dip cups to sterilize teats after milking
- Teat dip
- Stainless-steel collection buckets (electric milkers come with collection containers)
- Utensil cleaner
- Sanitizer or acid sanitizer
- Milk strainer
- Milk strainer filters
- Hand soap
- Paper towels
- Bottle brush
- Milk socks
- Extra rubber parts for equipment
- Bag balm

The girls make a beeline to the barn door. *Cheryl Kimball*

The does organize at the milking parlor entrance; they enter the milk parlor in about the same order every day. *Cheryl Kimball*

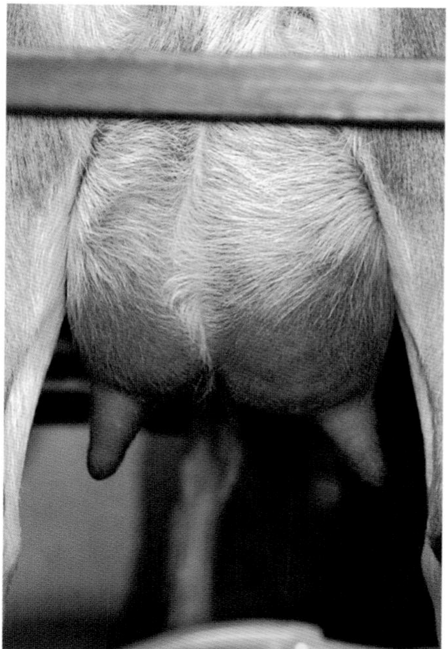

Teats are dipped in antiseptic solution before milking begins. *Cheryl Kimball*

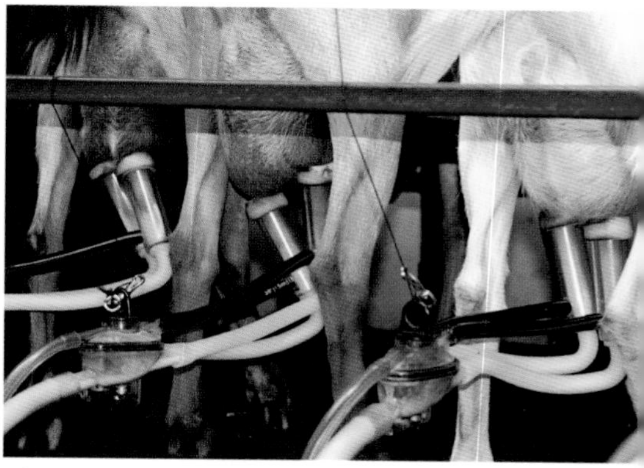

The milking machines are attached to the teats, and the milk parlor fills with the swooshing sound of the machines at work. *Cheryl Kimball*

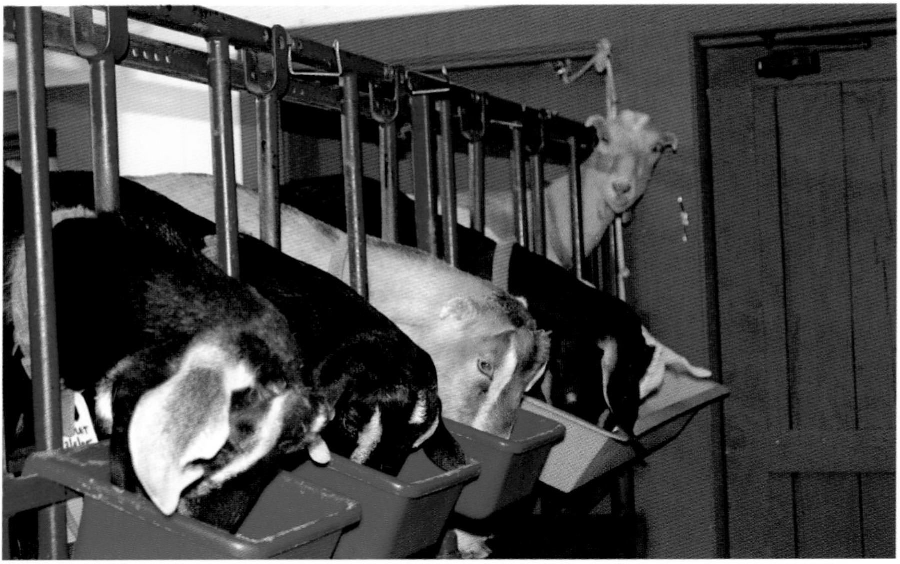

The does enjoy their ration of grain while they are being milked. *Cheryl Kimball*

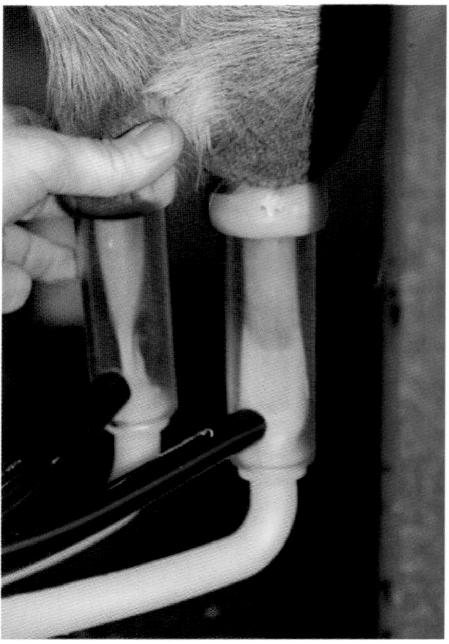

With machines, the milking process takes just a few minutes. *Cheryl Kimball*

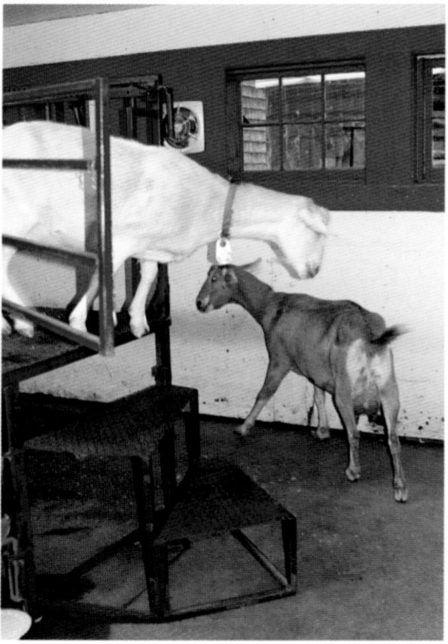

Six does are released to relax and eat hay for the night, and the next six are let into the milking parlor. *Cheryl Kimball*

Feeding

Feeding a dairy herd requires a bit more science than feeding a backyard herd of pet goats. Certain strong-flavored foods taint the milk, so you won't be as inclined to feed dairy goats scraps like you might pet goats. To maximize milk production, you need to give the dairy goats high-quality feed. Dairy does need to get the right nutrition to produce milk and maintain their own health. And you need to figure out the right quantity of feed for your breed of goat, for individual differences, and for the lifestyle your goats lead.

Hay

Hay is the most important food source you will give your dairy goats. Milking goats require high-quality, highly palatable alfalfa hay as a fundamental of their diet. This legume is highly nutritious, loaded with calcium (a reason NOT to feed it to male goats, who commonly have problems with urinary calculi buildup), and goats just love it.

Grain

To keep milk production up and to be sure your goats get the nutrients they need for their own maintenance, you will probably need to add grain to your feeding program. Commercial feed producers almost all have a line of goat feed that includes a mix designed for milking animals. Feed by weight and age and whether the goat is pregnant or not, and at what stage of pregnancy.

Fresh Food

If you have the space to turn out your goats on fresh pasture and browse, by all means let them at it! You do, however, need to be aware of poisonous plants (see chapter 2). Most livestock are pretty good about avoiding harmful plants. In their usual fashion, goats will take a taste and then move on, typically not ingesting enough to cause harm. Many toxic plants are unpleasant and even bitter tasting, which helps prevent a second tasting. Some plants, like yew,

Goats usually avoid bitter-tasting poisonous plants, but it's still best to check and clear your pasture of any potentially harmful growth. *Nicole Riley*

often used as an ornamental, are lethal in even small amounts—just a mouthful can destroy major organs. Some plants are only toxic in certain forms. Just a few red maple leaves, for example, can be lethal—but only when withered and dry. Underfed animals on depleted pastures may eat things they normally wouldn't touch. Sometimes livestock develop a taste for a plant that is toxic to them, so take precautions and check your pasture carefully and regularly.

If your milking herd has significant pasture space and can move around quite a lot throughout the course of a day, they may need some supplemental feed to make up for the energy used during that activity.

Don't worry about having to round up your animals from the pasture come milking time. If you keep a tight milking schedule and feed them their grain while they are being milked, the goats will be lined up at the milk parlor door waiting for you.

Milk Side Products

Goat milk doesn't have to be used only for drinking: It's a natural for a base of lotions and soaps. It is gentle to the skin and rich in natural moisturizing ingredients such as proteins, lipids, vitamins, and minerals. It is also close to the normal pH of skin and therefore considered less harsh than regular soap.

Many businesses specialize in goat milk products. Jenness Farm in Nottingham, New Hampshire, has made a good business from its goat milk products. It has a newly expanded retail store and sells everything from natural bug repellent to, of course, a huge selection of scented goat milk soaps and lotions. Jenness Farm is home to several Arapawa goats, a rare breed that only has a few hundred animals existing. The farm keeps the milk from its Arapawa does separate and makes soap from it, a portion of whose proceeds go to help the breed.

Lactose:
A sugar present in milk that is less water soluble than other kinds of sugars (such as glucose). It is digested by the enzyme lactase, which is in abundance in most of us as children but whose presence often decreases as we mature. Goat milk contains significantly less lactose than cow milk and human milk, so lactose-intolerant children are often fed goat milk.

The natural moisturizing properties in goat milk means it makes great soap and lotion products. *Cheryl Kimball*

The Nigerian Dwarf is one of the eight dairy breeds recognized by the ADGA. Although smaller than its standard-sized counterparts, the Nigerian produces competitive amounts of high-butterfat milk every day. *Daniel Johnson*

Meat

A byproduct of the goat dairy industry, buckling kids are typically raised to market weight and sold for meat. The rare individual is kept for breeding purposes, but between the need for high-quality individuals that have the best characteristics and the need to avoid inbreeding, not many male dairy kids make it to this position. When an exceptional one comes along that can't be used in the herd, the buckling can be sold to another, unrelated herd.

Milking does may also be culled from the dairy herd and find their way to the meat market. Candidates include does who are not breeding quality, whose milk production has decreased because of age, or who are proving to have difficult pregnancies.

You can choose to sell meat from the farm or sell it at a farmer's market or to a local small market. Goat meat is most commonly used in Middle Eastern and Spanish cuisine. It is flavorful, naturally lean, and is often baked, barbecued, or stewed and cooked with a variety of strong spices. Goat meat is referred to as *cabrito* from very young goats and *chevon* from older goats. The American Meat Goat Association has a recipe book for goat meat on its website (www.meatgoats.com).

The Meat Goats of Smoke Ridge

Montana ranchers Craig Tucker and Yvonne Zweede-Tucker have been raising meat goats at Smoke Ridge for nearly twenty years. Their summer herd (including the annual crop of kids) comprises nearly one thousand goats, mainly Spanish crossed with Boer or Savannah: "We want to keep the survivability, longevity, and fantastic maternal traits of the Spanish goat while adding some additional muscling."

The herd not only provides a product for the meat market come fall, it serves as brush control during the summer months. The goats browse on tough weeds and brush that cattle ignore in favor of more tender grasses: "The goats will walk (or rather, run!) through belly-deep grasses to demolish a wild rose bush and will consume knapweed flowerheads with gusto." The herd is fattened for market exclusively on the protein-rich weeds and brush, with no grain supplement.

Central Montana's open pastures are the perfect place for raising meat goats, but the demand for *chevon* lies hundreds of miles away, in the Midwest and on the East and West coasts. To keep costs under control, Smoke Ridge partners with neighboring goat ranches to ship animals to the buyers: "This way, we can take a full load of 200 animals on a 24-foot double-decked gooseneck trailer to the Pacific Northwest, or fill a quad-decked semi with up to 700 goats for the buyers."

Yvonne Zweede-Tucker, Smoke Ridge

The Boer is a popular meat goat, bred for fast weight gain and heavy muscling. *Copper Creek Boers, Oregon*

To create a goat meat operation, you will need to decide whether to butcher the animals yourself or hire it out to a slaughterhouse. Be sure to pick a reliable and well-run slaughter operation, since this is a critical step in the goat-to-meat process.

Getting USDA approval is a rigorous process, so be sure to find out all the details before you commit to starting a meat goat operation.

Fiber

Yet another goat product is fiber. There are two kinds of fiber goats: Cashmere goats, which produce the luxury fiber of the same name, and Angora goats, which produce mohair. The Cashmere goat is not a breed but just a goat that had been cultivated to produce a commercial-quality double fleece; the longer outer coat consists of coarse guard hairs and the undercoat consists of the fine valuable fibers used for wool.

The fleece is the wool of a fiber goat taken in a single shearing. Good fleece is typically characterized by its staple and crimp. Staple is the length of individual hairs; crimp is the kinkiness of the clumps of hairs. The fleece of Angora goats grows about an inch per month, and the goats are typically shorn twice a year.

Fiber production is a good job for wethers. They don't have reproduction stresses to take into consideration, and they can just enjoy themselves while eating and growing their fleeces! Many factors go into fleece production, including the overall health of the animal, genetic stock, and nutritional maintenance.

Fleece can be sold raw by the pound or it can be "processed," a term used for fleece that has been carded (fibers teased apart with a comblike instrument), washed, and often spun. Processed wool is ready to be used for knitting, weaving, or other fiber arts.

Although cashmere and mohair are expensive fleece products, huge amounts of money are not likely in this business due to the cost of maintaining the herd. But a fleece-producing goat herd can easily pay for itself.

Adult Angoras produce luxurious mohair yarn for knitting and weaving. *Daniel Johnson*

Land Clearing

While goats can be a huge help in clearing brush, you need one of two things for this to be successful:

- A big herd of goats
- A way to fence a small herd of goats into a concentrated area

Goats are browsers. They eat a little, move on, eat a little more, try this branch and that bush and that weed, move on. Goats on pasture never stay in one place long enough to make a dent. So if you have just a few goats and think you are going to use them to clear a few acres, think again!

Woven electric rolled fencing with a portable battery or solar fence charger can be used to confine your goats in an area until they have cleared it to your satisfaction. Then you can move them to another area. It can be a time-consuming way to clear land, but it is low impact, that's for sure; it's also a way to feed goats for free. If you choose to use the rotating fencing method, you must make certain the goats have access to fresh, clean water at all times.

Some people have tethered goats and tried to confine them to an area that way, but this method can be fraught with problems. Goats are not very good about keeping themselves untangled. They can easily choke themselves, or they can get tangled into one spot in the hot sun.

Herds of goats have been used to keep power lines clear of brush. *Shutterstock*

Showing Goats

Most agricultural fairs include a goat show. There are several good reasons to exhibit goats, including the following:

- **Children's projects**. It's a great goal for children to learn to care for livestock. Goats are a manageable size for children to handle; children and goats generally get along very well. Goats make a good 4-H project for school-age children.
- **Show credits**. If you are breeding or using goats for almost any purpose—milk, meat, fiber—it can be beneficial for your foundation stock to have show credits. Selling breeding rights to a "Grand Champion" buck or selling offspring from a proven superior milk-producing doe can mean a real monetary difference.
- **Farmer interaction**. Attending shows gives the goat farmer the chance to interact with other goat farmers and exchange ideas and information.
- **Fun**. Showing goats can simply be personally rewarding and fun.

Like all livestock shows, goats are judged based on a set of standards for their breed, which is then broken down by age and sex. A doe is judged on her udder size (the bigger the udder, the bigger the production potential) and teat size and placement (teats should enhance, not inhibit, nursing and milking). Bucks are judged on overall conformation. All breeds are also judged on that breed's specific standards of physical characteristics like ear type, overall size, and coloring.

As with dog shows, all goat champions in the individual classes compete at the end of the show to become the Grand Champion—the animal who best represents any of the breeds.

The ADGA sanctions shows for the eight recognized dairy breeds: Alpine, Oberhasli, Saanen, Nigerian Dwarf, Toggenburg, LaMancha, Nubian, and Sable. However, there are usually many other classes available to enter at a goat show, such as "Grade Doe," which is an animal that has the representative characteristics of a breed but is not registerable, as well as "Dam and Daughter," and "Best Udder."

Handlers usually enter the show ring and circle together around the ring, showing the judge their animals moving. Then they line up for individual visual and hands-on inspections. The judge then begins to sort out his or her final picks, sometimes asking that two goats be walked side by side for a final decision, at which point the judge lines the contestants up in order of final placing.

Getting Ready for the Show Ring

Showmanship is an important aspect of showing any kind of livestock. You need to know how to move the animal around the show ring and how to present its best attributes to the judge. Your animal must work with you while being led and shown; an uncooperative show animal does not get high marks from the judge. Even if your goat has a great udder and is well proportioned from a conformational standpoint, it will be hard for a judge to determine that if your goat is standing on its hind legs or racing around you, bleating furiously in the ring.

There are many opportunities to learn goat showmanship: the USDA, 4-H, the FFA Organization, and "goat camps" (often sponsored by a 4-H group) all have showing classes.

The American Goat Society (www.americangoatsociety.com) offers tips to those new at goat showmanship:

- Choose an animal that complements your size and that you can handle with ease.
- Make sure your goat is accustomed to being led and standing in "show stance" for judging: squared up with all four feet under the animal.
- Groom, shampoo, and trim your goat appropriately before the show.
- Know the basics about your goat—her age, her number of freshenings, her milk production.
- Choose a collar that doesn't draw attention away from your fine goat.
- Know how to move around the show ring appropriately.
- Be on deck ready and waiting to be called into the show ring.
- When you enter the ring, walk slowly and calmly and walk on the opposite side of the judge; never walk on the side between the judge and your animal.
- Smile and always watch the judge.
- Learn from your experience how to do a better job next time.

Another aspect of good showmanship is making sure your goat is buffed up and spit-shined and looking her best. Shampoo your goat with a livestock-safe shampoo. Brush her and clip her according to the standards of the breed. Typically this means trimming a doe's beard and perhaps even a full or partial body clip. Clip the udder area with a #30 blade. Dairy bucks are often trimmed all over except the head, leaving the beard and facial hair.

The judge of a goat show ranks your animal with a scorecard. See the sample USDA scorecard to get an idea of what the judge looks for in a show animal.

A champion LaMancha doe is exhibited at a goat show. *Susan Stewart*

Sample Goat Show Scorecard

The following scorecard is used to judge dairy goats at ADGA-sanctioned shows. *American Dairy Association*

	Points	Senior Doe	Junior Doe	Buck
A.	**GENERAL APPEARANCE** An attractive framework with femininity (masculinity in bucks), strength, upstandingness, length, and smoothness of blending throughout that create an impressive style and graceful walk.	35	55	55
	Stature—slightly taller at withers than at hips with long bone pattern throughout.	2	2	2
	Head & Breed Characteristics—clean-cut and balanced in length, width, and depth; broad muzzle with full nostrils; well-sculpted, alert eyes; strong jaw with angular lean junction to throat; appropriate size, color, ears, and nose to meet breed standard.	5	10	8
	Front End Assembly—prominent withers arched to point of shoulder with shoulder blade, point of shoulder, and point of elbow set tightly and smoothly against the chest wall both while at rest and in motion; deep and wide into chest floor with moderate strength of brisket.	5	8	10
	Back—strong and straight with well-defined vertebrae throughout and slightly uphill to withers; level chine with full crops into a straight, wide loin; wide hips smoothly set and level with back; strong rump which is uniformly wide and nearly level from hips to pinbones	8	12	10

	Points	Senior Doe	Junior Doe	Buck
	and thurl to thurl; thurls set two-thirds of the distance from hips to pinbones; well defined and wide pinbones set slightly above and smoothly set between pinbones; tail symmetrical to body and free from coarseness; vulva normal in size and shape in females (normal sheath and testes in males).			
	Legs, Pasterns & Feet—bone flat and strong throughout leading to smooth, free motion; front legs with clean knees, straight, wide apart and squarely placed; rear legs wide apart and straight from the rear and well angulated in side profile through the stifle to cleanly molded hocks, nearly perpendicular from hock to B, yet flexible pastern of medium length; strong feet with tight toes, pointed directly forward; deep heels with sole nearly uniform in depth from tie to heel.	15	23	25
B.	**DAIRY CHARACTER** Angularity and general openness with strong yet refined and clean bone structure, showing freedom from coarseness and with evidence of milking ability giving due regard to stage of lactation (of breeding season in bucks). **Neck**—long, lean, and blending smoothly into the shoulders; clean-cut throat nd brisket. **Withers**—prominant and wedgeshaped with the dorsal process arising slightly above the shoulder blades. **Thighs**—in side profile, moderately incurving from pinbone to stifle; from the rear, clean and wide apart,	8	12	10

	Points	Senior Doe	Junior Doe	Buck
	highly arched and out-curving into the escutcheon to provide ample room for the udder and its attachement. **Skin**–thin, loose, and pliable with soft, lustrous hair.			
	Legs, Pasterns & Feet –bone flat and strong throughout leading to smooth, free motion; front legs with clean knees, straight, wide apart and squarely placed; rear legs wide apart and straight from the rear and well angulated in side profile through the stifle to cleanly molded hocks, nearly perpendicular from hock to B, yet flexible pastern of medium length; strong feet with tight toes, pointed directly forward; deep heels with sole nearly uniform in depth from tie to heal.	20	20	30
C.	**BODY CAPACITY** Relatively large in proportion in size, age, and period of lactation of animal (of breeding season for bucks), providing ample capacity, strength, and vigor.	35	55	55
	Chest–deep and wide, yet clean-cut, with well sprung foreribs, full in crops and at point of elbow.	4	4	7
	Barrel–strongly supported, long, deep, and wide; depth and spring of rib tending to increase into a deep yet refined flank.	6	8	8
	Mammary System–Strongly attached, elastic, well-balanced with adequate capacity, quality, ease of milking, and indicating heavy milk production over a long period of usefulness.	35	-	-

	Points	Senior Doe	Junior Doe	Buck
	Udder Support—strong medial suspensory ligament that clearly defines the udder halves, contributes to desirable shape and capacity, and holds the entire udder snugly to the body and well above the hocks. Fore, rear, and lateral attachments must be strong and smooth.	13	-	-
	Fore Udder—wide and full to the side and extending moderately forward without excess non-lactating tissue and indicating capacity, desirable shape, and productivity.	5	-	-
	Rear Udder—capacious, high, wide, and arched into the escutcheon; uniformity wide and deep to the floor; moderately curved in side profile without protruding beyond the vulva.	7	-	-
	Balanced, Symmetry & Quality—in side profile, one-third of the capacity visible in front of the leg, one-third under the leg, and one-third behind the leg; well-rounded with soft, pliable, and elastic texture that is well collapsed after milking, free of scar tissue, with halves evenly balanced.	6	-	-
	Teats—uniform size and of medium length and diameter in proportion to capacity of udder, cylindrical in shape, pointed nearly straight down or slightly forward, and situated two-thirds of the distance from the medial suspensory ligament on the floor of each udder-half to the side, indicating ease of milking.	4	-	-

Other Uses for a Goat

People have found all manner of reasons to have goats around. Meat, milk, and fiber are the most common. Here are some of the less common ones.

Petting Zoo

Goats are great animals to include in a petting zoo designed to teach children about farm animals. They tend to be friendly, and if they are raised well, they won't bite or be rude or engage in dangerous behavior. Children like to feed them, which can be fine, but make sure that they aren't overfed on high-calorie grain and snacks, or you will be dealing with sick goats. And goats can be so food-motivated that feeding them on a reward system can make them quite aggressive and pesty, and they can quickly become too dangerous for children.

Cart Pulling

Some breeds of goats, especially wethers, can grow quite large. They are very strong animals and can pull a cart holding a child or small adult with ease. Carts, sleighs, harnesses, and even a goat-powered garden cultivator can be found at Hoegger Goat Supply (see chapter 1).

Pack Animals

Goats make great hiking and camping companions. They can carry packs and are so people-oriented that they love "coming along." Goats are such great climbers that they will readily follow you up inclines and over rocks. If you are going to hike overnight with them, you will need to make sure the goat is protected from predators. Buy packs made for goat use. Bring water and food for them for during transport (they can carry their own as

Trained pack goats carry provisions over a mountain trail for a camping trip. *Scott Herbolsheimer, Summit Pack Goat*

well as yours). In most environments, the pack goats are capable of living off the land once you've reached your camping destination.

Be sure to adequately prepare your goats for any lengthy or overly strenuous trip. Like any animal, they cannot be taken out of the pen where they have lounged for two years and suddenly go on a strenuous hike. They need to build up to it! Keep in mind the differences between a pampered pet and a trained pack goat accustomed to two-week-long camping trips.

This goat looks like he told himself a good joke. *Shutterstock*

Energy Generators

Goats are easily trained, and they can be convinced to keep a treadmill going to power small farm equipment such as corn huskers.

Companion Animals

Goats have traditionally served as companions to racehorses, which cannot be turned out with other horses for fear of injury. A goat keeps the horse company in its long days confined to a stall. These companion goats are typically wethers, but nonmilking does can be used as well.

Sheer Entertainment

And last, a small collection of goats is good to have around for sheer entertainment value! Just keep your setup simple and feed them well, and these hardy animals will cheer you up for years to come. Pygmies and dwarfs may seem like the best choice because they are small and they eat less, but remember that they do require the same amount of care. They also can be difficult to fence in; they aren't as strong as full-sized goats, but they can get through much smaller openings!

Horses and goats naturally get along and make great companions. *Shutterstock*

References

Belanger, Jerry. *Raising Milk Goats the Modern Way*. North Adams, MA: Storey Publishing, 1990.

Haynes, N. Bruce. *Keeping Livestock Healthy*. North Adams, MA: Storey Books, 2001.

McCurnin, Dennis M. and Joanna M. Bassert. *Clinical Textbook for Veterinary Technicians*. 5th ed. Philadelphia: W. B. Saunders Company, 2002.

Merck Veterinary Manual. 9th ed. Whitehouse Station, NJ: Merck & Company, Inc., 2005.

Healthy goats make happy charges. *Shutterstock*

Glossary

abomasum. This is the fourth stomach chamber of the ruminant. The abomasum is considered the true stomach, since this is where enzymatic digestion occurs. This chamber is most similar to the stomach of nonruminants such as humans.

abscess. A localized collection of pus, somewhat common in goats. Also known as a *boil*.

afterbirth. (See *placenta*.)

alopecia. Hair loss. Many parasites, skin conditions, and allergies cause hair loss. It is usually accompanied by itching.

caseous. Resembling cheese or curd.

dystocia. Difficulty with giving birth. This is something dairy goat farmers quickly become familiar with. Although goats generally give birth with ease and with no human intervention necessary, if you have enough goats giving birth, you will have a couple of dystocias to attend to.

enterotoxemia. A disease caused by toxins building up in the digestive tract. Also known as "overeating disease."

eruct. To burp. Goats do a lot of eructating.

herpes. A class of virus. Goats are susceptible to orf, a form of the herpes virus. (See *orf*.)

hypocalcemia. Abnormally low blood calcium. Calcium problems need to be watched for in dairy herds and in breeding animals.

ketosis. A metabolic disease seen in ruminants that is caused by incomplete metabolism of fatty acids. It is usually caused by carbohydrate deficiency, and sometimes through malnutrition or starvation, which causes the animal to start using its own body stores for energy. It can also be seen in animals that have high-fat diets or diabetes.

lactate. To secrete milk. Dairy animals are "lactating." Milk contains the protein "lactose," although goat's milk is lower in lactose than cow's milk.

mastectomy. Removal of the udder. This can be necessary due to a tumor.

metritis. Inflammation of the uterus. Metritis can be a problem following parturition if the placenta is retained or if bacteria get into the uterus.

necropsy. The veterinary term analogous to the human term "autopsy." This is an examination of the body after death, to determine the cause of death. In certain states, livestock necropsies may be required by law when the death may have involved a contagious disease that could threaten livestock in general.

omasum. The fourth stomach chamber of the ruminant.

parasite. An organism that lives in (*internal parasites*) or on (*external parasites*) another organism. The nonparasite is known as the "host." Goats are prone to many worms and lice and other parasites.

parturition. The act of giving birth.

placenta. The saclike structure in the uterus that holds the fetus and allows it to receive nourishment for growth. The placenta is expelled from the uterus at birth or shortly thereafter. Many animals eat the placenta (also known as "afterbirth") to avoid attracting predators. A placenta that does not expel from the uterus is problematic; the animal should be seen by a veterinarian as soon as possible or this will become an emergency, as the retained placenta decomposes and becomes toxic to the animal.

pyometra. Pus accumulated due to infection in the uterine cavity.

rumen. The first and main stomach compartment of a cud-chewing animal, or "ruminant."

ruminant. Goats, sheep, and cattle are all ruminants: They ingest food into the first of their four stomach compartments, known as the rumen, where it begins the digestion process. Part of the process is for the animal to regurgitate "cuds" of food from the rumen, rechew it, and swallow it again.

scours. Severe diarrhea in farm animals. Many young livestock experience a bout with scours not long after birth. If not carefully monitored, scours can leave a young animal seriously dehydrated.

vector. A carrier, such as an insect, of disease from one infected animal to another.

withdrawal time. The interval between the administration of a drug and the time that the animal can legally be slaughtered for consumption or that its milk can be sold. This is an important concept for the livestock farmer.

A Pygmy goat enjoys fresh hay, which is essential for confined goats. *Shutterstock*

Index

Housing for goats can be simple and small, so long as it provides protection from the elements. *Shutterstock*

The Field Guide to Cattle
9780760331927

The Field Guide to Rabbits
9780760331934

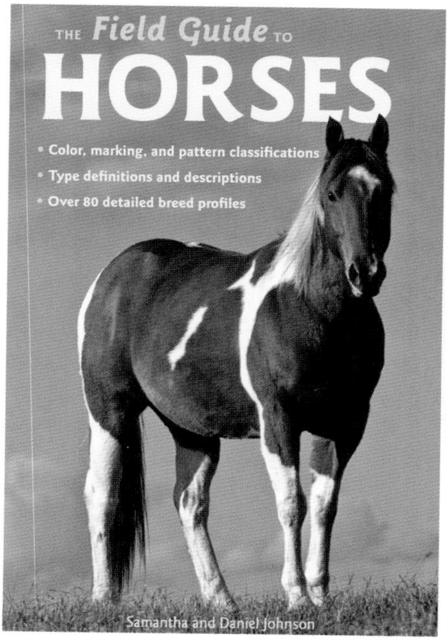

The Field Guide to Horses
9780760335086

The Field Guide to Chickens
9780760324738

About the Author

Cheryl Kimball is a certified veterinary technician and the author of several books, including *The Complete Horse* and *Mindful Horsemanship*. She works part time at the New Hampshire Farm Museum in Milton and lives in Middleton, New Hampshire, with her husband, dogs, cats, horses, and goat.

Courtesy of Jack Savage